W0107648

Toward a Molecular Basis of Alcohol Use and Abuse

Edited by
B. Jansson, H. Jörnvall, U. Rydberg, L. Terenius
Karolinska Institutet, Stockholm, Sweden
B.L. Vallee
Harvard Medical School, Boston, MA, USA

1994. 406 pages. Hardcover. ISBN 3-7643-2940-8 (EXS 71)

The chapters in this volume consider subjects ranging from genetics, markers, and molecular biology of alcoholism, to clinical observations and treatment. The aim is to integrate pertinent information from the fields of molecular and cell biology with clinical and pharmacological knowledge in the attempt to establish molecular bases of alcohol use and abuse. An introductory section summarizes historical aspects of alcohol use, and subsequent chapters concern novel drugs, pharmacological aspects, gene structures, cloning, and enzymatic properties.

Several "non-traditional" alcohol scientists number among the authors; their contributions highlight parallels and the possibility of interaction between different fields.

Novel results of particular interest include up-dated summaries on receptors, enzymes, and other proteins, as well as corresponding gene structures and regulation, setting the basis for distinguishing markers and pinpointing further possible pharmacological treatments.

Please order through your bookseller
or directly from:
Birkhäuser Verlag AG, P.O. Box 133
CH-4010 Basel / Switzerland
Fax ++41 / 61 721 79 50
E-Mail: 100010.23@compuserve.com

Orders from the USA or Canada
should be sent to:
Birkhäuser Boston, 333 Meadowlands Parkway,
Secaucus, NJ 07094-2491 / USA
Call Toll-Free 1-800-777-4643

For more information on recent and forthcoming books and journals you can order the BIRKHÄUSER LIFE SCIENCES BULLETIN, published twice a year and free of charge.

Springer Basel AG

International Symposium on Nicotine: The Effects of Nicotine on Biological Systems II

Satellite Symposium of the XIIth International Congress of Pharmacology, Montreal, Canada, July 21–24, 1994

The Abstracts

Edited by
P. B. S. Clarke
M. Quik
K. Thurau
F. Adlkofer

Springer Basel AG

Editors:

Professor Paul Brian Sydenham Clarke
Department of Pharmacology and Therapeutics
McGill University
3655 Drummond Street
Montreal
Québec
Canada H3G 1Y6

Professor Maryka Quik
Department of Pharmacology and Therapeutics
McGill University
3655 Drummond Street
Montreal
Québec
Canada H3G 1Y6

Professor Dr. Dr.h.c. Klaus Thurau
Physiologisches Institut
Universität München
Pettenkoferstrasse 12
D-80336 München
Germany

Professor Dr. Franz Adlkofer
Stiftung VERUM
Stiftung für Verhalten und Umwelt
Pettenkoferstrasse 12
D-80336 München
Germany

CIP catalogue records for this book are available from the Library of Congress,
Washington D.C., USA and Deutsche Bibliothek Cataloging-in-Publication Data.

Originally published by Birkhäuser Verlag in 1994

Camera-ready copy prepared by the authors
Printed on acid-free paper produced from chlorine-free pulp

ISBN 978-3-7643-5087-1   ISBN 978-3-0348-7416-8 (eBook)
DOI 10.1007/978-3-0348-7416-8

9 8 7 6 5 4 3 2 1

Contents

Oral Presentations

Session 1: Overview

CURRENT CONTROVERSIES IN NICOTINE RESEARCH. Paul B.S. Clarke, Dept. of
Pharmacology and Therapeutics, McGill University, Montréal, Québec, Canada H3G
1Y6

S 1

Nicotine research has proliferated alarmingly in recent years and the profusion of data
makes it hard to maintain a critical eye on developments outside one's immediate
speciality. I intend to take aim at a number of widely–held views which I believe
(perhaps mistakenly) deserve to be questioned. I will touch upon the following issues:
Is nicotinic cholinergic neurotransmission an established feature of the mammalian
nervous system ? Does blockade of CNS nicotinic receptors result in cognitive (or
other) deficits ? Do neurons that express nicotinic receptors precisely control where
these are located ? Do nicotinic receptors containing α4 and ß2 subunits represent a
particularly prevalent population in brain and do they mediate pharmacological actions
of smoking doses of nicotine to any important extent ? In studies where animals are
treated repeatedly with nicotine, are the drug effects observed due to the acute effects of
chronic treatment or the chronic effects of acute treatment ? Is the upregulation of ^{3}H–
nicotine binding sites in brain that is seen upon chronic in vivo nicotine treatment
triggered by changes in the functional status of these receptors ? Are the *central*
actions of nicotine important in tobacco smoking ? Many habitual smokers appear
dependent on nicotine, so why then does nicotine replacement therapy help only a small
minority to remain abstinent in the long–term ? (answers to all the above to be provided
by conference participants !).

ACUTE BIOLOGICAL EFFECTS OF NICOTINE AND ITS METABOLITES. Neal L. Benowitz, M.D.
University of California, San Francisco.

S 2

Nicotine is the primary psychoactive chemical in tobacco and is the main cause of tobacco
addiction. In addition to its psychoactive effects, nicotine acts on other physiological systems that
contribute to addiction and potentially to adverse health consequences of tobacco addiction. Cotinine,
the major metabolite of nicotine, may also contribute to the biological effects of nicotine. This paper
will review recent findings on the biological effects of nicotine and cotinine with emphasis on effects
in humans. Issues to be discussed include the following. Rapid delivery of nicotine from tobacco
smoke or nicotine replacement products produces more intense physiological and psychological effects
than does nicotine delivered more slowly. Hemodynamic effects of nicotine include sympathetic
nervous system activation, which may include constriction of coronary arteries. The cardiovascular
effects of cigarette smoking and transdermal nicotine differ, the latter resulting in a less sympathetic
neural activation and little or no activation of platelets or coagulation. Nicotine increases metabolic
rate, both by sympathetic nervous system activation and by enhancement of lipolysis with futile
cycling, and also suppresses appetite by a central mechanism. The result is lower body weight in
smokers', which returns to normal after cessation of smoking. Nicotine increases hepatic metabolism
of free fatty acids, an effect that is expected to lead to a more atherogenic lipid profile. Tolerance
develops to the biological effects of nicotine, but the extent and the rate of tolerance differ for
different effects. High and low level cigarette smokers differ in responsiveness to nicotine for some
effects (like the increase in metabolic rate), but not in other effects (such as heart rate acceleration).
Cotinine appears to have psychoactivity and may modify symptoms of nicotine withdrawal. In
animals, cotinine affects a number of enzymes including those involved in the synthesis of estrogen,
testosterone and aldosterone. The significance of these actions in humans is as yet unclear.

INVOLVEMENT OF NICOTINE AND ITS METABOLITES IN THE PATHOLOGY OF SMOKING
AND SMOKING-RELATED DISEASES: FACTS AND HYPOTHESES. F. X. Adlkofer. VerUm,
Stiftung für Verhalten und Umwelt, Pettenkoferetraße 12, D-80336, Munich.

S 3

In Germany, as in several other countries, smoking has become synonymous with nicotine
abuse suggesting that it is nicotine which is causally implicated as a risk factor in the
etiopathology of smoking-related diseases. More than 150 years of nicotine research, since
the alkaloid was first isolated by Posselt and Reimann in 1828, have not succeeded in
elucidating all the pharmacological effects of this drug. Today it is generally accepted that
nicotine, at smoking-related concentrations, binds mainly to receptors at autonomic ganglia,
the adrenal medulla, neuromuscular junctions and in the brain thus leading to the release of
neurotransmitters, such as acetylcholine, epinephrine, norepinephrine, dopamine and
serotonin. The question arises as to how nicotine, through these or other mechanisms, may
contribute to the pathogenesis of smoking-associated cancer and respiratory and
cardiovascular diseases, and to what extent established scientific knowledge supports the
numerous hypotheses on this issue.

Session 2: Structure and function of nicotinic receptors

S 4

ASSEMBLY OF THE ALPHA7 CONTAINING NEURONAL NICOTINIC ACETYLCHOLINE RECEPTOR. Jim Patrick, David Char, DaNong Chen, Lorna Colquhoun, Hong Dang, Finn Goldner, Santosh Helekar, Kelly Dineley, and Shawn Neff. Division of Neuroscience, Baylor College of Medicine, Houston, Texas 77030.

The neuronal nicotinic acetylcholine receptor gene family encodes proteins that form either homo-oligomeric or hetero-oligomeric ligand gated ion channels. Although the total number of combinations of possible receptors is large, only a few have been demonstrated to function in the oocyte expression system. However, those that have been expressed differ from each other in several important ways. Receptors expressed in oocytes differ with respect to the ligands that activate them, their single channel properties, their permselectivity, and their modulation by external calcium. Furthermore, some cells express several of the genes that encode these receptor subunits and assemble receptors that contain a subset of the expressed genes. There are, therefore, probably mechanisms responsible for the assembly and targeting of receptors containing specific combinations of subunits. We investigated this possibility initially focussing on the homo-oligomeric alpha7 receptor and identified a requirement for the activity of a peptidyl prolyl isomerase in the assembly of this receptor. This requirement is shared by the serotonin-gated ion channel. We have created a number of mutations in the alpha 7 protein to try to identify relevant proline residues and have examined the expression of these mutant receptors in the *Xenopus* oocyte expression system.

S 5

BIOCHEMICAL CHARACTERIZATION OF NEURONAL NICOTINIC RECEPTORS. C. Gotti, F. Clementi. CNR Center of Cytopharmacology, Department of Medical Pharmacology, University of Milano, 20129 Milano, Italy
Neuronal nicotinic acetylcholine receptors (nAChRs) belong to the ligand-gated ion channel family and are found throughout the peripheral and central nervous system. Recent studies have demonstrated substantial diversity in neuronal nAChR structure, physiology and pharmacology and this is due to the presence of different subtypes of nAChR with different pharmacological and electrophysiological properties. The distribution of different nAChR subunits in monkey and human brain is discrete, selective for particular nuclei and is different among species. Among the nAChR subtypes present in the nervous system, the most elusive in terms of structure and function is still the nAChR which binds and/or is blocked by α-Bungarotoxin (αBgtx). As a model system to study the structure and function of the neuronal αBgtx receptors we have employed chick brain and the human neuroblastoma cell lines. At least two subunits have been cloned that bind αBgtx, α7 and α8. We have recently purified and functionally reconstituted αBgtxRs present in chick optic lobe (COL) and cerebellum. When reconstituted in a planar lipid bilayer after agonist binding, these receptors from a cation channels, the opening of which is blocked by cholinergic antagonists. Using polyclonal antibodies against synthetic peptides of chick α7 and α8 αBgtx receptor subunits we have immunopurified and characterized the α7, α7-α8 subtypes present in chick optic lobe and retina. We have studied their biochemical and pharmacological properties and characterized their biophysical properties after reconstitution in lipid bilayer. By using these Abs specific for the cloned chick α7 and α8 subunits we found that in chick brain and retina there is a family of αBgtx receptors composed of at least three members: the α7, the α8 and the α7-α8 subtypes. With the same approach we have also studied the biochemical and biophysical properties of human αBgtxR in neuroblastoma cell lines.

S 6

NEURONAL NICOTINIC RECEPTOR STRUCTURE AND FUNCTION Jon M. Lindstrom Ph.D., Department of Neuroscience, Medical School of the University of Pennsylvania, Philadelphia, PA 19104-6074
There are three branches of the neuronal nicotinic receptor gene family: 1) muscle receptors, 2) neuronal receptors which do not bind αbungarotoxin, and 3) neuronal receptors which do bind αbungarotoxin. In mammalian brain the primary subtype of neuronal nicotinic receptor which does not bind αbungarotoxin has the subunit composition $(\alpha4)_2(\beta2)_3$. These receptors have high affinity for nicotine. Mouse fibroblasts transfected with these receptors exhibit a nicotine-induced increase in receptor amount similar to that observed in smokers or animals chronically exposed to nicotine. Upregulation results from a nicotine-induced decrease in the rate of receptor destruction in these fibroblasts. It seems likely that nicotine-induced upregulation of receptor amount does not require ion flow through the receptor, because the non-competitive antagonist mecamylamine also causes upregulation. In brain the primary subtype of neuronal nicotinic receptor which can bind αbungarotoxin contains α7 subunits, perhaps in combination with unknown subunits. In chick retina, αbungarotoxin-binding receptors containing α8 subunits predominate, and a subtype containing both α7 and α8 subunits is a minority component in both brain and retina. α7 and α8 receptors differ in pharmacological properties, with α8 receptors having lower affinity for αbungarotoxin and higher affinity for small cholinergic ligands than do α7 receptors. Both α7 and α8 homomers expressed from cDNAs in *Xenopus* oocytes have similar channel properties. Their high permeability for Ca^{++}, rapid desensitization, and strong inward rectification suggest that α7 and α8 receptors may participate in unusual synaptic mechanisms.

DETERMINANTS REGULATING NEURONAL NICOTINIC RECEPTOR FUNCTION
D. Bertrand, S. Bertrand, I. Forster and J.-P. Changeux Dpt of Physiology, Geneva, Dpt of Physiology, Zürich, Switzerland and Institut Pasteur, Paris, France.

To evaluate amino-acid determinants in the structure function relationship of the neuronal nicotinic receptors (nAChRs), we used site directed mutagenesis in combination with electrophysiological recordings. The homomeric chick $\alpha 7$ receptor represents a unique tool for this approach and was used throughout this work. Channels reconstituted with the $\alpha 7$ subunit display a high Ca^{2+} permeability comparable to that of NMDA receptors. Point mutations in the $\alpha 7$ second transmembrane domain (TM2) have revealed the existence of two independent sites that control this permeability. Furthermore, we observed that, as other neuronal nAChRs, the $\alpha 7$ channels are allosterically modulated by the extracellular calcium concentration. Moreover, substitution of the amino acid E327 concomitantly abolishes the voltage dependency of these receptors, indicating that this charged ring is involved in the inward rectification. We found that internal Mg^{2+} acts as an open channel blocker, somewhat similar to that observed with the external Mg^{2+} in the NMDA receptors.

EXPRESSION, FUNCTION, AND REGULATION OF NEURONAL ACH RECEPTORS CONTAINING THE $\alpha 7$ GENE PRODUCT. D. Berg, W. Conroy, R. Corriveau, P. Pugh, M. Rathouz, S. Romano, S. Vijayaraghavan, and Z.-w. Zhang. Dept. of Biology, Univ. of Calif., San Diego; La Jolla, CA 92093.

Nicotinic acetylcholine receptors (AChRs) that bind α-bungarotoxin (αBgt) are widely distributed throughout the nervous system, and until recently remained a mystery with respect to function. Chick ciliary ganglion neurons contain large numbers of such receptors (αBgt-AChRs) located almost exclusively in nonsynaptic membrane. Of the 10 known neuronal AChR gene products, only the $\alpha 7$ is found in most (>95%) ciliary ganglion αBgt-AChRs. Whole cell patch clamp techniques together with rapid application of agonist demonstrate that the receptors function as ligand-gated ion channels that are cation-selective, prefer nicotine over ACh, rapidly desensitize, and elevate intracellular calcium. This last feature raises the possibility that αBgt-AChRs utilize calcium as a second messenger to set in motion a cascade of cellular events. One consequence of calcium modulation by the receptors may be morphogenic remodeling of neurite arborization. The receptors are located on cell processes both in vivo and in culture. Activation of the receptors in culture induces neurite retraction in a calcium-dependent manner, acting in part through voltage-gated calcium channels. A second consequence of αBgt-AChR activation is the calcium-dependent release of the membrane-permeant second messenger arachidonic acid. Arachidonic acid, in turn, can reversibly inhibit the receptors, providing an opportunity for negative feedback regulation. The ability of αBgt-AChRs to elevate intracellular calcium both directly and through activation of voltage-gated calcium channels may also mediate cholinergic regulation of cellular events in non-neuronal cells. The $\alpha 7$ gene is expressed in developing muscle at early stages and produces an $\alpha 7$-containing complex devoid of $\alpha 1$ subunits but having the size of a fully assembled receptor. The functional consequences of $\alpha 7$-containing receptors in muscle have yet to be examined. (Supported by NS12601, NS25916, and TRDRP)

REGULATION OF ACETYLCHOLINE RECEPTOR GENES EXPRESSION DURING SYNAPTOGENESIS IN MUSCLE AND BRAIN. Jean-Pierre CHANGEUX, J.L. Bessereau, A. Bessis, A. Duclert, C. Le Poupon, H.O. Nghiêm, A.M. Salmon, N. Savatier. CNRS UA D1284 "Neurobiologie Moléculaire", Institut Pasteur, 25 rue du Dr Roux, 75724 Paris Cedex 15, France.

At the adult motor endplate, the AChR protein [$\alpha 2 \beta \gamma / \epsilon \delta$] as its subunits mRNAs are localized exclusively under the motor nerve ending. Denervation of the adult muscle causes a reappearance of unspliced and mature mRNA in extra-junctional areas. On the other hand, in the non innervated embryonic myotube, the α, β, γ and δ subunit mRNA are distributed all over the cell. A compartmentalization of gene expression at the level of subneural "fundamental" nuclei therefore takes place during development and is analyzed by the methods of cell biology and recombinant DNA technology (transfection, transgenic mice, DNA injection, adenoviral vectors) with both cultured and developing muscles in situ.

The data are interpreted in terms of a model which assumes that: 1) in the adult muscle fiber, nuclei may exist in different states of gene expression in subneural and extrajunctional areas; 2) different second messengers elicited by neural factors such as CGRP (cAMP) or ARIA (tyrosine kinase) (under the nerve endings) or electrical activity (Ca^{++}, protein kinase C) (outside the endplate) regulate the state of transcription of these nuclei via trans-acting allosteric proteins binding to distinct cis-acting DNA regulatory elements; 3) in the chicken α-subunit enhancer, consensus E Boxes (CANNTG) play a differential role in the regulation by electrical activity while in mouse ϵ-subunit promoter a different 83 nucleotides sequence confers preferential synaptic expression; 4) a regulation of the expression of myogenic protein genes takes place during endplate formation; 5) multiple post-transcriptional processes involving, in particular, the Golgi apparatus, proteins from the basal lamina and from the cytoskeleton (e.g. the 43KD protein) contribute to the clustering, and stabilization of the AChR in the postsynaptic membrane. A similar "promoter approach" is extended to the analysis of the expression of brain nicotinic receptor genes $\alpha 2$, $\alpha 3$, $\alpha 4$, $\beta 2$, $\beta 4$ by in situ hybridization, transfection of cultured nerve cells ($\beta 2$-subunit promoter) and transgenesis in mice (chick $\alpha 2$ subunit gene and mouse $\beta 2$-subunit promoter).

Session 3: Nicotinic receptor regulation and tolerance

S 10   BIOCHEMICAL MEASURES OF NICOTINIC RECEPTOR DESENSITIZATION. M.J.Marks, S.R.Grady, S.F.Robinson, A.E.Bullock, and A.C.Collins. Institute for Behavioral Genetics, University of Colorado, Boulder, CO.

The properties of desensitization of nicotinic receptors in synaptosomal preparations isolated from mouse brain were examined using biochemical assays. While the efflux of $^{86}Rb^+$ and the release of [^{3}H]dopamine from synaptosomes can be stimulated by nicotine and other nicotinic agonists, and inhibited by nicotinic antagonists, the pharmacological properties of these two processes differ. Prolonged exposure of synaptosomes to activating concentrations of nicotine (0.1 μM - 10 μM) desensitizes both thalamic $^{86}Rb^+$ efflux and striatal [^{3}H]dopamine release. Desensitization can also be observed when the synaptosomes are exposed to nonactivating concentrations of nicotine (1 nM - 100 nM). The kinetics of desensitization of the functional responses are in many ways similar to the kinetics of high affinity [^{3}H]nicotine binding. Both the functional desensitization and the ligand binding are adequately, but not perfectly, explained by the two state model of Katz and Thesleff. The desensitization of $^{86}Rb^+$ efflux from thalamic synaptosomes varies with the agonist such that the rate of desensitization observed for acetylcholine and other quarternary agonists exceeds that for nicotine or cytisine. The $^{86}Rb^+$ efflux stimulated by cytisine was substantially lower than that observed for the other agonists. The nicotine-stimulated $^{86}Rb^+$ efflux is inhibited approximately 50% by the sodium channel antagonist, tetrodotoxin (TTX). The rate of desensitization is reduced somewhat in the presence of 100 nM TTX, but the EC_{50} for activation by nicotine is unaffected by this toxin. Inclusion of Cs^+ or other K^+ channel blockers has no effect on either the amount of efflux or the rate of desensitization. Although full responsiveness for nicotine-stimulated [^{3}H]dopamine release recovers when the agonist is removed, the $^{86}Rb^+$ efflux after desensitization recovers incompletely. These studies demonstrate that nicotinic receptors isolated from brain undergo functional desensitization. Such desensitization may be important in regulating in vivo response to both acute or chronic exposure to nicotine or other agonists. (Supported by DA-03194 and DA-00194).

S 11   THE ROLE OF DESENSITIZATION IN CNS NICOTINIC RECEPTOR FUNCTION. P.M. Lippiello, M. Bencherif and R.J. Prince*. R&D Department, R.J. Reynolds Tobacco Co., Winston-Salem, NC 27102 and *Integrated Toxicology Program, Duke University, Durham, NC 27710

The kinetics of ligand binding to rat brain nicotinic receptors can be described by a two-state model whereby nicotinic agonists can bind to either of two pre-existing receptor conformations, one presumed to be activatable (low affinity) and the other desensitized (high affinity). Kinetic constants describing rates of association, dissociation and ligand-induced conformational changes were estimated from [^{3}H]-nicotine binding. These parameters were used to predict functional effects of nicotine in vitro, using [^{3}H]-dopamine release from rat brain striatal synaptosomes. In general, the model was a good predictor of receptor activation and desensitization by nicotine. In order to determine the relationship between high affinity receptor binding and receptor activation, receptor binding and dopamine release were determined for some representative nicotinic agonists. There was no correlation between binding affinity and either EC_{50} or Emax for dopamine release. To examine the relationship of ligand structure to receptor desensitization, the effects of tetramethylammonium iodide (TMA) were studied. Although less potent than nicotine, TMA was found to be a full agonist for dopamine release and was blockable by mecamylamine. The EC_{50}'s for desensitization of dopamine release compared well with affinity constants determined from receptor binding; however, the rate and extent of desensitization by TMA and nicotine were essentially the same. These results suggest that high affinity nicotine receptor binding is a good predictor of relative potency for functional desensitization rather than activation and that the positive charge center of nicotinic ligands is a sufficient structural feature for both activation and desensitization.

S 12   **PRESYNAPTIC HETERORECEPTORS, AUTORECEPTORS AND NICOTINIC RECEPTOR SUBTYPES**

S. Wonnacott. School of Biology and Biochemistry, University of Bath, BATH BA2 7AY UK

Molecular cloning strategies have shown that numerous nicotinic receptor (nAChR) subunits are expressed in the vertebrate nervous system, and the potential for distinct nAChR subtypes is enormous[1]. The subunit combinations that constitute native nAChRs are, in most cases, uncertain, as is the identity of nAChRs mediating different functions in the CNS. In the absence of subtype-specific ligands, we have compared the pharmacological profiles of nAChRs. Nicotinic autoreceptors mediating [^{3}H]ACh release from hippocampal synaptosomes, nicotinic heteroreceptors mediating [^{3}H]dopamine release from synaptosomes prepared from striatum and frontal cortex, and α4β2 nAChR stably expressed in M10 cells[2] have been examined. Nicotine-evoked [^{3}H]ACh release gives sharply bell-shaped dose-response curves, in contrast to [^{3}H]dopamine release. Closer examination of striatal dopamine release evoked by nicotine indicates a biphasic dose-response curve, consistent with heterogeneity of presynaptic heteroreceptors. While α4β2 nAChR are compatible with the autoreceptor mediating [^{3}H]ACh release, this comparison is not definitive, and more subtype-selective ligands are urgently needed.

1. Sargent, P. (1993) Ann. Rev. Neurosci. 16: 403-433.
2. Whiting, P. et al. (1991) Mol. Pharmacol. 40: 463-467

REGULATION OF NEURONAL NICOTINIC RECEPTORS: *IN VIVO* **AND** *IN VITRO* **STUDIES.**

K.J. Kellar, M.I. Dávila-García, Y. Xiao, R.A. Houghtling, R.D. Mellon, S.S. Qasba and C.M. Flores. Department of Pharmacology, Georgetown University, Washington, DC 20007.

In rat brain, the predominant nicotinic receptor that displays high affinity (nM) binding sites for nicotinic cholinergic agonists is comprised of α4 and ß2 subunits. Chronic exposure to agonists such as nicotine or cytisine increases binding to these receptors *in vivo*. This increase is not blocked by co-administration of the nicotinic receptor channel blocker mecamylamine, suggesting that the mechanism underlying the increase in receptor binding sites may be triggered by simple agonist occupation of the receptor and may be independent of the actual function of the receptor. Concurrent with this increase in binding sites, the response to nicotine *in vivo* is markedly diminished. Studies of the mechanisms involved in the regulation of these receptors have been hampered by the absence of a cell model. We have used embryonic day 18 rat neuronal cells in primary culture to begin to study this regulation. The binding sites for nicotinic receptor agonists in these cultured neurons have pharmacological characteristics similar to those in adult rat brain, including high affinity for [^{3}H]cytisine as well as for nicotine, carbachol and dihydro-ß-erythroidine. While [^{3}H]cytisine binding is measurable in primary cultures from several brain areas, including cerebral cortex, hippocampus and striatum, the binding is highest in cultures from mesencephalic neurons ($\approx$ 50 fmol/mg protein). Culturing primary cortical neurons in the presence of nicotine for seven days or longer, increases the number of nicotinic binding sites. These primary cultures should be useful as a model system for studies of the intracellular signal pathways and molecular mechanisms involved in the regulation of these neuronal nicotinic receptors, including studies of transcriptional events and receptor turnover studies.

CONDITIONED TOLERANCE TO NICOTINE IN RATS A. R. Caggiula, L. H. Epstein, S. M. Antelman, S. Knopf, K. A. Perkins, S. Saylor, E. Donny and R. Stiller Departments of Psychology, Psychiatry and Anesthesiology, University of Pittsburgh, Pittsburgh, PA 15260 and Pittsburgh Cancer Institute

Recent evidence from our laboratory indicates that the development of tolerance to several behavioral and endocrine effects of nicotine depends, at least in part, on learned associations with environmental cues that reliably signal drug delivery. Administration of nicotine without those cues partially or completely reinstates the original drug response; i.e., disrupts the expression of tolerance. Additional research from several laboratories has led to the unique hypothesis that this conditioned tolerance is mediated by activation of the hypothalamic-pituitary-adrenocortical (HPA) system. This paper will deal with evidence bearing on several critical assumptions upon which this hypothesis is based: (1) environmental cues (CS) -- by virtue of their repeated association with nicotine -- activate the hypothalamic-pituitary-adrenocortical (HPA) system; (2) HPA activation releases a hormone -- corticosterone -- which can reduce nicotine responsiveness acutely, within the time frame of the CS-UCR (unconditioned response) interval; (3) the CS-induced HPA activation, which is reflected in increased corticosterone, is the causal agent of reduced responsiveness to nicotine. Supported by DA07546

Session 4: Neuronal, trophic and endocrine effects of nicotine

S 15   DEVELOPMENTAL EFFECTS OF NICOTINE. T.A. Slotkin. Dept. of Pharmacology, Duke University Medical Center, Durham, NC 27710.

A growing body of evidence indicates that nicotine, acting at fetal CNS nicotinic cholinergic receptors, produces neurobehavioral teratogenesis. By contrasting acute nicotine injections to pregnant rats with continuous infusions (implanted minipumps), we have been able to separate the contributing variables of hypoxia/ischemia, maternal nutrition and fetal growth retardation from the direct effects of nicotine exerted on developing neurons. Nicotine leads to premature arrest of neural cell replication, an effect mimicking the natural actions of acetylcholine as a neurotrophic factor controlling differentiation of cholinergic target cells. With exogenous nicotine administration, this trophic effect is elicited at an incorrect time, leading to aberrant macromolecule synthesis and lasting changes in activity of a variety of neurotransmitter pathways. Because nicotinic receptors also play similar roles in peripheral pathways, we have been able to demonstrate clear-cut functional deficits in cardiac-sympathetic activity in animals exposed to nicotine prenatally, characterized by deficiencies in sympathetic tone, in β-adrenergic receptor binding sites and in heart rate responses to adrenergic agonists or electrical stimulation of sympathetic nerves. One important implication of neurotransmitter receptors as a target for trophic effects on the immature nervous system is that "classical" teratologic indices, such as growth retardation, may be inappropriate markers of developmental neurotoxicity, since receptor-mediated actions typically occur at drug doses orders of magnitude below those for non-selective toxic events. For nicotine, doses that do not interfere with maternal or fetal weight gain, and that do not evoke any signs of general toxicity, are still fully effective in perturbing CNS cell development and synaptic function.

(Supported by a grant from the Smokeless Tobacco Research Council).

S 16   POTENTIATION OF TRANSMISSION VIA PRESYNAPTIC NICOTINE-ACTIVATED CHANNELS PERMEABLE TO CALCIUM AND BLOCKER BY a-BgTx. L.W. Role. Dept. of Anatomy and Cell Biology, Columbia Univ., New York.

It is surprising that there are so few clear examples of synaptic transmission mediated directly via nicotinic ACh receptors. In contrast, there is growing evidence for the expression of nicotinic binding sites in terminal fields of serotonergic and dopaminergic neurons where the local infusion of nicotine has been shown to evoke transmitter release. We have monitored both evoked and spontaneous transmission at two identified synapses: projections of the autonomic preganglionic nucleus of Terni to the neurons of the lumbar sympathetic ganglia. These synapses were chosen based on: (a) the ability to selectively extirpate both the pre and post synaptic partners prior to the establishment of synaptic transmission (b) the ability to recapitulate reliable synaptic transmission *in vitro* (c) which, in turn, permits detailed biophysical analyses and controlled manipulation of nAChR subunit expression and most importantly (d) evidence that nAChR subunits and/or nAChRs are expressed by the presynaptic neurons and are present in their terminal fields. Nicotine potently and profoundly regulated spontaneous and evoked synaptic transmission at both synapses. Habenula-IPN synapses, mediated by released glutamate, were potentiated by direct application of nicotine. Analysis of effects of spontaneous synaptic currents indicates alterations in the release of glutamate without concomitant effects on the amplitude of unitary events, suggesting a selective effect of nicotine at presynaptic sites. Likewise, examination of ACh-mediated synaptic transmission in sympathetic ganglia revealed pronounced synaptic facilitation by nicotine at concentrations two orders of magnitude lower than those required to activate postsynaptic nAChRs. Activation of presynaptic nAChRs increases Ca in presynaptic neurites and is blocked by aBgTx. Furthermore, deletion of the aBgTx-sensitive synaptic facilitation. These studies indicate an important regulatory role for nicotine in the CNS, controlling transmitter release by a selective activation of presynaptic nAChRs that may include the $a7$ subunit.

S 17   **ELECTROPHYSIOLOGY OF NICOTINIC RECEPTORS IN RODENT CNS**
Clement Léna, Jean-Pierre Changeux and Christophe Mulle - Laboratoire de Neurobiologie Moléculaire, Institut Pasteur, Paris, France.

Although neuronal nicotinic AChR genes are widely expressed in the vertebrate CNS, their role in the function of synaptic circuits in the CNS remains elusive. To address this issue, we have directly analysed the functional properties of nicotinic receptors in various regions of the brain, including the medial habenula, the interpeduncular nucleus and the thalamus, in acutely isolated neurons or in slice preparations. At least three different types of nicotinic receptors can be distinguished that differ in their electrophysiological and pharmacological properties. High calcium permeability seems to be a common property of neuronal nicotinic receptors. In medial habenula neurons, calcium influx through nicotinic activated channels is physiologically significant and operates at membrane voltages where both voltage-gated channels and NMDA channels do not flux any calcium. In addition, the activity of these receptors is finely tuned by the external concentration of calcium. Despite the abundance of nicotinic receptors on perisomatic regions of many neurons, we have not succeeded to record nicotinic synaptic currents in these regions. However, we have shown that nicotinic agonists could trigger the release of GABA by acting on "preterminal" nicotinic receptors in a TTX-dependent manner. On the other hand, presynaptic nicotinic receptors localized on synaptic boutons could also influence neurotransmitter release in a TTX independent manner. These data point to the large functional diversity of nicotinic receptors in the rodent CNS.

FACTORS CONTROLLING NICOTINIC ACETYLCHOLINE RECEPTOR EXPRESSION ON RAT SYMPATHETIC NEURONS.

E. Cooper and P. De Koninck, Department of Physiology, McGill University, Montréal, Québec. H3G 1Y6.

Determining factors that control the expression of neurotransmitter receptors and the mechanisms by which these factors operate is essential to understand how synapses form during development and in the adult. Both innervation and activity play an important role in controlling the expression of neuronal nicotinic acetylcholine receptor (nAChR) subunits. In this presentation, we will discuss mechanisms by which activity can regulate the expression of nAChR subunits in neonatal rat sympathetic neurons. In addition, we will discuss some of the functional implications of changes in subunit mRNA levels on the properties of these receptors.

A ROLE FOR THE NICOTINIC α-BUNGAROTOXIN RECEPTOR IN GROWTH RELATED PROCESSES.

M. Quik. Dept. Pharmacol., McGill Univ., Montreal, Quebec, Canada.

Although molecular biology approaches have yielded a multiplicity of neuronal nicotinic receptor subunits (α2-9, ß2-4), electrophysiological and receptor binding studies have allowed for a somewhat simpler classification dependent on the ability of the neuronal receptors to interact with the snake toxin α-BGT. There is a general consensus that α-BGT insensitive nicotinic receptors may be involved in mediating cholinergic transmission. In contrast, the role of the α-BGT sensitive nicotinic receptor population has been more difficult to assess. Initially, correlative evidence had implicated these sites in developmental and/or trophic events. Such an interpretation is further supported by the results of recent studies in neuronal, as well as non-neuronal cells. Exposure of nerve growth factor treated PC12 cells (a model for neuronal type cells) to nicotine resulted in a decline in neurite outgrowth which could completely be reversed when cells were simultaneously incubated in the presence of nM concentrations of α-BGT; these observations suggest that activation of the nicotinic α-BGT receptor may modulate process outgrowth in these cells. α-BGT receptors have also been identified on small cell lung carcinoma cells, which are of neuroendocrine origin. Interestingly, nicotine and another nicotinic agonist cytisine resulted in an increase in cell number which could effectively be blocked by nM concentrations of α-BGT. These observations suggest that the nicotinic α-BGT receptor population may be involved in a growth related role in both neuronal and non-neuronal cells. This work was supported by the MRC Canada.

MECHANISMS OF NICOTINE STIMULATED CELL PROLIFERATION IN NORMAL AND NEOPLASTIC NEUROENDOCRINE LUNG CELLS. H.M. Schuller, Carcinogenesis and Developmental Therapeutics Program, University of Tennessee, Knoxville TN 37901, USA.
The secretion of biogenic amines and neuropeptides by normal pulmonary neuroendocrine (PNE)cells is under cholinergic control. Nicotine mimicS this effect via nicoti-nic acetylcholine receptor (nAChR) stimulation. Because many of these products are autocrine growth factors, we have addressed a potential mitogenic effect of nicotine on normal (fetal hamster) and neoplastic (human lung carcinoid and small cell cancer) PNE cells *in vitro*. Our data show that nicotine has a saturable mitogenic effect on normal and neoplastic PNE cells maintained in an atmosphere of high (10%) CO_2 while it fails to stimulate cell proliferation when the cells are maintained at 5% CO_2. Mitogenesis in response to nicotine is completely inhibited by the ganglionic nAChR blocker hexamethonium while antagonists of 5-HT receptors and PKC cause a significant but not in all cell types complete inhibition of the nicotinic stimulation. As the observed promoting effect of high CO_2 on cell stimulation by nicotine appears to mimic the documented stimulation of PNE cells in individuals with chronic lung disease , we studied the effects of different concentrations of this gas alone on cell proliferation. Our data show that CO_2 causes concentration-dependent and satu-rable mitogenesis in normal and neoplastic PNE cells which is inhibited by antagonists of 5-HT receptors and PKC but not by hexamethonium.We conclude that CO_2 and nicotine synergistically stimulate related mitogenic signal transduction pathways some of which include a serotonergic loop and activate PKC downstream. Our data are in accord with the documented higher risk of smokers with chronic obstructive lung disease for the development of lung cancer and suggest that nicotine contributes significantly to the carcinogenic burden associated with smoking. In support of this interpretation, nicotine induced a significant number of respiratory tract tumors in hamsters exposed simultaneously to hyperoxia, a condition resulting in the development of pulmonary fibrosis and elevated intrapulmonary CO_2 levels.

BRAINSTEM CATECHOLAMINERGIC PATHWAYS ACTIVATED BY NICOTINE ARE INVOLVED IN THE HIPPOCAMPAL EXPRESSION OF c-FOS mRNA and PROTEIN, STIMULATION OF THE HYPOTHALAMIC PARAVENTRICULAR NUCLEUS AND SECRETION OF ACTH. B. Sharp, Endocrine-Neurosci. Labs, Mpls Med Res Fndn, Depts. of Med, Hennepin County Med Ctr and U of Minn

Nicotine is a potent stimulus for the secretion of ACTH and the selective expression of c-fos mRNA and protein in the brain. Peak levels of c-fos mRNA, present 30-60 min after systemic nicotine, were inhibited by mecamylamine (Mec). Nicotine (0.05 mg/kg) increased cFos protein in these areas: catecholaminergic brainstems regions (NTS-A2 and -C2), hypothalamus [paraventricular (PVN) and supraoptic nuclei (SON), parvocellular and oxytocinergic regions; respectively], and limbic system [cingulate gyrus (CG), amygdala]. A higher dose of nicotine (0.1) further increased expression in A2, PVN, SON and CG. In addition, this dose induced cFos in locus coeruleus (LC) and dentate gyrus (DG). Propranolol pretreatment, but not spiperone, abolished expression in the DG, and reduced it in CG. In contrast, PVN was unaffected. Nicotine dose-dependently elevated plasma ACTH when administered into the fourth (IV) ventricle, and stimulation by i.v. nicotine was inhibited by delivering Mec into the IV ventricle. Microinjecting nicotine (50 nl) into brainstem catecholaminergic regions rapidly stimulated ACTH secretion; rank order of regional responsiveness was A2 > C2 > LC >>A1 (C1 did not respond). ACTH responses to IV ventricular nicotine depended on central catecholamines, inhibited by α-adrenergic antagonists in the hypothalamic III ventricle). Additionally, administration of nicotine systemically or into the IV ventricle stimulated norepinephrine release into the PVN, as determined by *in vivo* microdialysis. Systemic nicotine also (i) stimulated cFos protein expression in the subpopulation of PVN neurons synthesizing corticotrophin-releasing factor (CRF) and (ii) depleted CRF from its axons in the median eminence. In summary, nicotine activates ascending catecholaminergic pathways, acting at a site(s) proximate to their origin in the brainstem. Catecholamine release is involved in nicotine-stimulated ACTH secretion and in the expression of c-fos mRNA in rostral brain regions, e.g. DG and CG. The pattern of cFos expression indicates that activation of A2 and C2 correlate with stimulation of the PVN (and ACTH) and SON, whereas LC correlates with DG.

NICOTINE-INDUCED GENE EXPRESSION OF PROENKEPHALIN IN BOVINE CHROMAFFIN CELLS: V.Höllt*, X.Wang°, and B.Bacher°, °Institute of Physiology, University of Munich and *Institute of Pharmacology and Toxicology, Otto-von-Guericke University, Medical Faculty, Leipzigerstr.44, D-39120 Magdeburg, Germany

The induction of the proenkephalin gene by nicotine has been characterized in bovine adrenal medullary chromaffin cells. Nicotine (10 μM) caused an about 4-fold increase in the proenkephalin mRNA levels within 24 hours.

The half-life of the proenkephalin mRNA in nicotine-stimulated cells was similar as that in control cells (about 13 hours) indicating that nicotine acts at the levels of gene transcription. This was directly demonstrated by showing that the expression of a proenkephalin chloramphenicol acetyl transferase (CAT) reporter gene (PENKCAT-153/+50) containing 153 nucleotides of upstream promoter sequences was increased (about 2-fold) by nicotine after transient transfection in the chromaffin cells.

In addition, nicotine induced a rapid, but transient elevation of the immediate early gene mRNAs c-fos. c-jun and jun-B. Maximally increased levels for c-fos mRNA (about 100-fold) were obtained after 20 min, for c-jun (3-fold) and jun-B mRNA (5-fold) 60 min after nicotine addition.

The expression of the proenkephalin gene reporter plasmid which contains a dimer of the ENKCRE-2 element in front of a minimal promoter was increased by co-transfection of a c-fos expression plasmid indicating that nicotine may induce the proenkephalin gene in chromaffin cells via c-Fos which binds to the CRE-2 element.

EFFECTS OF NICOTINE ON THROMBOXANE AND LEUKOTRIENE SYNTHESIS IN CELLULAR SYSTEMS
M. Goerig and S. Koll Med. Klinik IV. Univ. of Erlangen, Germany.

Nicotine has been shown to be a potent inhibitor of thromboxane formation in myelomonocytic cells and is able to decrease prostacyclin formation of endothelial cells. To investigate mechanisms of nicotine-mediated effects on eicosanoid synthesis of the vessel wall we used coculture systems of confluent human iliac venous or arterial endothelial cells (HIVEC/HIAEX) and human myelomonocytic cells. Our experiments have shown that: 1. Thromboxane synthesis is synergistically increased in cocultures of monocytic cells with endothelial cells by a factor of at least 35 when compared with monocultured cells. This effect is cell contact-mediated and depends on a selective upregulation of the key enzyme cyclooxygenase II in endothelial cells. In contrast, prostacyclin synthesis is decreased.

2. 10 to 100 nm nicotine is an effective inhibitor of transcellular thromboxane synthesis in cocultures of endothelial cells and mono-cytic cells. Nicotine directly inhibits the activity of thromboxane synthase and prevents the upregulation of cyclooxygenase II in endothelial cells.

3. 10 to 100 nm nicotine potentiates leukotriene B_4 and leukotriene C_4 synthesis in cocultures of endothelial cells with monocytic cells and in cocultures with granulocytic cells. These results indicate that in vivo relatively low concentrations of nicotine may dramatically change transcellular synthesis of eicosanoids and may favor leukotriene formation in conditions of vascular inflammation.

SMOKING–INDUCED ALTERATIONS IN BRAIN ELECTRICAL ACTIVITY: NORMALIZATION OR ENHANCEMENT? V. Knott. Dept. of Psychiatry, University of Ottawa and Institute of Mental Health Research/Royal Ottawa Hospital, Ottawa, Ontario, Canada, K1Z 7K4.

S 24

One of the attractions of cigarette smoking may lie in nicotine's ability to enhance normal levels of functioning and/or normalize deficient levels of functioning which may characterize transient situational states or enduring maladaptive traits. Brain electrical recordings from the scalp surface provide a non–invasive strategy for examining the impact of smoking on central functions and their relationship to perceptual, cognitive, and emotional processes. Quantitative electroencephalography (QEEG) and event–related potentials (ERPs) collected during passive and task-activated conditions were employed here to examine putative normalizing/enhancing effects of acute smoking in smokers relative to non–smoking non–smokers. Tasks were specifically selected to tap attentional and memory processes, processes which have been previously reported to be affected by smoking and nicotine administration in both normal and pathological populations. Evidence indicates that the ability of smoking to affect brain function is response/task dependent and that evidence for both normalization and enhancement are observed with neuroelectric recordings.

NICOTINE AND COGNITIVE EFFECTS
I. Hindmarch. Human Psychopharmacology Research Unit, University of Surrey, Milford Hospital, Godalming, GU7 1UF, United Kingdom.

S 25

The most satisfactory explanations of tobacco smoking have proposed that smokers are influenced by the psychoactive effects of nicotine. Research conducted at the HPRU suggests that improvements in human cognition and psychomotor performance after the administration of nicotine may be central to the habitual use of tobacco. Investigations into the effects of nicotine on critical flicker fusion threshold show that where the background level of central nervous system arousal is sub-optimal (particularly among abstinent smokers), nicotine acts as a mild cognitive enhancer; this in turn may be the basis of improved performance found in studies investigating the effects of nicotine on attention, particularly the reduction of the vigilance decrement, and is consistent with a cholinergic mode of action for nicotine. Sensori-motor and memory function are also improved by nicotine, and these effects are found in non-abstinent smokers and non-smokers. This may reflect a different mechanism of action, possibly through the release of dopamine. In the wider context, it appears that smokers may be able to manipulate nicotine intake and other parameters of smoking to control and optimize their cognitive and psychomotor performance. If this is shown to be the case, it could follow that nicotine also has a role in modulating mood and assisting in the amelioration of acute and chronic psychological distress. In conclusion, the psychopharmacological profile of nicotine is one of small, but reproducible, specific positive effects on human cognition and psychomotor performance.

EVIDENCE THAT NICOTINE IS ADDICTIVE. I.P. Stolerman and M.J. Jarvis. Section of Behavioural Pharmacology and ICRF Health Behaviour Unit, Institute of Psychiatry, De Crespigny Park, London SE5 8AF, UK.

S 26

Despite the wide-ranging and authoritative 1988 review by the US Surgeon General, views questioning the addictiveness of nicotine continue to be expressed in some quarters. This lack of complete consensus is not unexpected, since no generally agreed scientific definition of addiction exists. However, several lines of evidence from both the human and animal literature support the view that nicotine is addictive, both according to the scientific criteria commonly applied to other drugs and according to the generally understood meaning of the word. Patterns of use by smokers and the remarkable intractability of the smoking habit point to compulsive use as the norm. Studies in both animal and human subjects have shown that nicotine can function as a reinforcer, albeit under a more limited range of conditions than with some other drugs of abuse. In drug discrimination paradigms there is some cross-generalisation between nicotine on the one hand, and amphetamine and cocaine on the other. Both tolerance and sensitization to effects of nicotine can occur, under different circumstances. A well-defined nicotine withdrawal syndrome has been delineated, which is alleviated by nicotine replacement. Nicotine replacement also enhances outcomes in smoking cessation, roughly doubling success rates. In total the evidence clearly identifies nicotine as a powerful drug of addiction, comparable to heroin, cocaine and alcohol. Addiction to each of these drugs has some unique features, and nicotine is no exception; enhancements of cognitive function are not incompatible with addictiveness but rather, they may constitute one of the mechanisms that sustain it. Thus, the notion that nicotine is 'habituating' but not addicting may be consigned to the archives of history.

S 27   SELF-ADMINISTERED NICOTINE ACTS THROUGH THE VENTRAL TEGMENTAL AREA -
IMPLICATIONS FOR DRUG REINFORCEMENT MECHANISMS. William A. Corrigall. Addiction
Research Foundation and Dept. of Physiology, University of Toronto, Toronto, ONT, Canada M5S 2S1.

Self-administration of nicotine depends on dopaminergic processes, and in particular on the mesolimbic
dopamine projection from the ventral tegmental area (VTA) to the nucleus accumbens. We have begun
to investigate how this interaction occurs. To determine the site of nicotine's action within this system,
we have used the nicotinic antagonist dihydro-ß-erythroidine (DHßE). Focal administration of DHßE into
the nucleus accumbens does not alter nicotine self-administration, but microinfusions of the antagonist
within the VTA produce a dose-related decrease in self-administration behavior. Similar intra-VTA
infusions of the antagonist do not affect motor output, and do not change operant behavior maintained
by food, or intravenous self-administration of cocaine. These observations show that self-administered
nicotine acts within the VTA to sustain voluntary drug-seeking behavior. In addition, the findings imply
that cholinergic mechanisms may influence dopaminergic reinforcement processes in the VTA. The
absence of an effect of DHßE on reinforced behavior maintained by food or intravenous cocaine suggests
that a tonically-active nicotinic input is not involved in reinforcement. Not surprisingly, muscarinic
antagonists do not alter nicotine self-administration or locomotor activity when infused into the VTA.
However, in animals with no nicotine exposure, muscarinic antagonists are effective after intra-VTA
infusion: spontaneous locomotor activity is increased, and reinforced behavior maintained by intravenous
cocaine or food is decreased, by intra-VTA infusions of scopolamine. These data suggest that there may
be a balance between nicotinic and muscarinic influences on VTA dopamine cells which can be shifted
towards a predominantly nicotinic influence during nicotine exposure.

S 28   DESENSITISATION OF THE STIMULATORY EFFECTS OF NICOTINE ON DOPAMINE
SECRETION IN THE MESOLIMBIC SYSTEM OF THE RAT D.J.K. Balfour and M.E.M.
Benwell Dept of Pharmacology, University Medical School, Ninewells Hospital, Dundee, DD1 9SY
Scotland.

It has been suggested that stimulation of mesolimbic dopamine (DA)-secreting neurones is an
important factor in the development of nicotine dependence. However, there is evidence that many
central nicotinic receptors are desensitised by chronic nicotine. The primary objective of the present
study was to establish the nicotine concentrations which caused desensitisation of mesolimbic DA
responses to the drug. Osmotic minipumps were used to constantly infuse nicotine or saline for 14
days. In addition the rats were given 5 daily sc injections of saline or nicotine prior to the test day.
Microdialysis studies, performed on day 14 showed that, in animals infused with saline, 5 daily sc
injections of nicotine increased (P<0.05) the peak DA overflow in the NAc (measured as percent of
baseline levels) evoked by an injection of nicotine from 113 ± 21 to 236 ± 81 %. The enhanced
responses were abolished in animals constantly infused with nicotine at doses of 4 and 1mg/kg/day
and attenuated in animals infused with 0.25mg/kg/day. These infusions yielded plasma nicotine levels
of 113 ± 21, 24 ± 5 and 9 ± 4ng/ml respectively. These and other supporting data suggest that the
receptors which mediate the stimulatory effects of nicotine on mesolimbic DA neurones are
desensitised by nicotine levels commonly found in the plasma of tobacco smokers and imply that
tobacco smoking may not invariably result in stimulation of mesolimbic DA neurones.

This study was supported by Forschungsrat Rauchen und Gesundheit and The Wellcome Trust

S 29   MECHANISMS OF ACUTE AND CHRONIC TOLERANCE TO THE BEHAVIORAL EFFECTS OF
NICOTINE. J.A. Rosecrans, J.R. James, L.D. Karan. Dept. Pharmacology &
Toxicology and Division of Substance Abuse Medicine, Va. Commonwealth
Univ., Richmond, VA 23298, USA.

Research conducted in this laboratory over the last 25 years has
focused on central mechanisms of nicotine action using a two-lever
operant drug-induced Discriminative Stimulus (DS) paradigm. The DS
procedure relies on the ability of an experimental subject (rat or
human) to be able to detect nicotine from vehicle in order to receive a
positive reward. This paradigm is especially interesting since rats do
not readily develop tolerance to its DS effects even though rats exhibit
pharmacological/behavioral tolerance to nicotine-induced (0.4-0.8 mg/kg,
s.c.) disruption of operant behavior; Fixed Ratio (FR) or Variable
Interval (VI) schedules of reinforcement. While rats trained to detect
nicotine (DS) do appear to develop tolerance to its disruptive effects,
recent work has also shown that a select population of rats are capable
of exhibiting acute tolerance to the nicotine DS. Experiments were
conducted utilizing drug discrimination and operant disruption paradigms
in order to learn more about these divergent effects. The results of
this work suggests that nicotine-induced tolerance (Acute or Chronic) is
contingent on the ability nicotine to induce a desensitization of select
central nicotinic-acetylcholinergic-receptors (nAChRs). (This research
was supported by a generous grant from the German Research Council on
Smoking and Health, Berlin, Germany).

BEHAVIORAL AND BIOCHEMICAL ANALYSIS OF DEPENDENCE PROPERTIES OF NICOTINE.

T.Yanagita, Y.Wakasa and K.Ando Preclinical Research Division, Central Institute for Experimental Animals, 1433 Nogawa, Miyamae-ku, Kawasaki 216 Japan

1) In intravenous self-administration experiments on nicotine in rhesus monkeys, studies of the relationships between infusion speed and intake rate and between size of unit dose and reinforcing effect were performed. In the former study the dose of nicotine was always 30 μg/kg but infusion speeds of 0.3, 1.3, and 5.2 μg/sec were used, with higher infusion speeds resulting in higher intake rates. Since the peak plasma level of nicotine paralleled the speed and since reinforcing efficacy is known to parallel the unit dose size within a certain dose range, the increase of intake rate was attributed to elevation of the nicotine level. In the latter study, self-administration was observed at 0.6, 2.5, 10, and 40 μg/kg for 2 to 4 weeks each dose level in both ascending and descending orders. As a result, the minimum unit dose to reinforce intake was found to be 10 μg/kg when tested in ascending order and 2.5 μg when tested in descending order. When nicotine 10 μg/kg was taken the plasma level was 7.8 ng/ml. Based on these findings, the minimum nicotine content in a cigarette to retain reinforcing effect in human smokers will be discussed.

2) The discriminative effects of nicotine infused into the brain were studied in rats. Rats were trained to discriminate a subcutaneous dose of nicotine 0.5 mg/kg from saline. Then, guide cannulae were implanted bilaterally into the brain. The effects of the subcutaneous nicotine were substituted by infusion of nicotine 40 μg/rat at the medial prefrontal cortex and 100 μg/rat at the nucleus accumbens. The effects were partially substituted by nicotine 20-60 μg/rat at the ventral tegmental area but not substituted by infusions at the dorsal hippocampus or medial habenular nucleus. Thus, the median prefrontal cortex and nucleus accumbens were regarded to be important sites of action in producing the subjective effects of nicotine which may result in its reinforcing effect and psychic dependence potential.

NICOTINE INTAKE IS REGULATED IN HUMANS M.A.H. Russell. Health Behaviour Unit,

Institute of Psychiatry and Maudsley Hospital, London SE5 8AF, United Kingdom.

That smokers alter their pattern of puffing and inhalaton and thereby regulate their nicotine intake independently of other smoke components to maintain their customary blood, and presumably brain, levels has been regarded as critical evidence for nicotine's central role in controlling the behaviour. This is now well established by a variety of approaches which include the down-regulation of intake from smoking when nicotine is administered from another source, its upregulation to give higher blood nicotine levels when the effects are blocked by mecamylamine, and its regulation either way to correct for alterations in urinary excretion induced by manipulations of urinary pH. Where uncertainties arise is with the precision of the regulation. Up-regulation to maintain usual blood nicotine levels is seldom complete and smokers tolerate and adjust quite quickly to a drop to about two-thirds of their usual peak smoking level. Although earlier studies indicated down-regulation to avoid exceeding usual levels to be fairly precise, more recent studies involving ad libitum smoking during prolonged infusions or while wearing transdermal nicotine patches indicate that smokers tolerate substantially higher peak and trough blood nicotine levels to maintain the average 10ng/ml boost per cigarette in venous blood which is characteristic of their usual smoking. This boost in inhaling smokers is taken as ten 0.1 mg puff by puff doses absorbed sufficiently rapidly to deliver high nicotine boli in arterial blood having transient concentrations of 100-150 ng/ml. Since at least 50% of smokers smoke every hour or more and start within 30 min of waking most of their nicotinic receptors may be desensitised. In those with prominent peaks after each cigarette the boli may activate some non-desensitised receptors, whereas those with less prominent peaks and boli may be motivated to maintain a desensitised state.

THERE IS MORE TO SMOKING THAN THE CNS EFFECTS OF NICOTINE

J. Rose VA Medical Center, Durham, NC 27705, Department of Psychiatry, Duke University, Durham, NC,

A great deal of evidence has implicated nicotine as a key reinforcing constituent in tobacco. However, it is often ignored that from a smoker's point of view, cigarette smoking involves much more than simply delivering nicotine to the brain. Simply put, nicotine without smoke may be as unsatisfying as smoke without nicotine. Three main lines of evidence supporting this conclusion are reviewed, from studies using: 1) nicotine administration through routes other than smoking; 2) comparison of responses to nicotine-containing and de-nicotinized cigarettes; and 3) effects of nicotine blockade with mecamylamine. These studies lead to the conclusion that conditioned reinforcing effects of sensory and motor aspects of smoking are important determinants of smokers' immediate subjective reaction to cigarettes; although nicotine delivery is important in the long-term maintenance of cigarette smoking, the perceived CNS effects of nicotine in the doses delivered from cigarette smoke are often subtle.

S 33   PHARMACOLOGICAL DETERMINANTS OF CIGARETTE SMOKING. J.E. Henningfield. Clinical Pharmacology Branch, Addiction Research Center, National Institute on Drug Abuse, Baltimore, MD, 21224, U.S.A.

For centuries, if not millennia, tobacco smoking was used ritualistically to produce intoxication and even hallucinations. European explorers of the New World discovered that once a person began to use tobacco, he might go to great lengths to continue its use. Since the turn of the twentieth century it has also been understood that these effects were due to the nicotine delivered by tobacco smoking, that nicotine is a potent and powerful effector of nervous tissue, and that tolerance to its effects could quickly develop. Observations over the past few decades have also confirmed that, for many persons, the difficulty in abstaining from tobacco is comparable that experienced by heroin, cocaine and alcohol dependent persons. Over the past three decades or so, major advances have been made in identifying the specific determinants of compulsive cigarette smoking. These include the development of tolerance and physiological dependence to nicotine, pharmacologic effects of nicotine that smokers consider important if not indispensable to their lives, and the pleasurable and reinforcing effects of smoking provided by the central and peripheral effects of nicotine and other smoke constituents. These actions are not mutually exclusive but their importance appears to vary across individuals. Qualitative and quantitative aspects of these effects are determined by the nicotine dosage form and level.

S 34   PSYCHOLOGICAL RESOURCES FROM NICOTINE
David M. Warburton, Dept. of Psychology, Reading University, Reading, RG6 2AL, United Kingdom.
Smokers state that they choose to smoke for the psychological benefits which they derive from the product. A large body of evidence shows that they obtain improved mood states and enhanced information processing capacity from smoking and that these effects can be related to the uptake of nicotine. The mood states consist of mild relaxing effects and mild pleasurable effects. Improved information processing capacity is seen in a variety of attentional tasks from simple vigilance to complex perceptual intrusion tasks. The use of complex paradigms, like semantic processing, show that complex processing can be improved by nicotine.

Experiment with a variety of nicotine delivery systems establish that the changes can be attributed to nicotine. Studies with minimal abstinence will be presented in order to demonstrate that these improvements do not depend on abstinence from nicotine.

These results can be interpreted in terms of a functional model of nicotine use. In this model, nicotine use can be seen as purposive, a behavior to obtain psychological resources. It can result from both exogenous and endogenous causes, the characteristics of the situation and of the individual.

S 35   NICOTINE IS ADDICTIVE. R. West Dept. of Psychology, St George's Hospital Medical School, London SW17 0RE.
Debate over whether nicotine is addictive can easily deteriorate into an argument over semantics. Yet the issue is far from academic because the way that society, health organisations and the legal system deal with smokers and those suffering from smoking-related diseases is critically determined by how far the behaviour is viewed as an illness that one should attempt to treat or the result of a free choice. It is noteworthy that in the debate there is little dispute among scientists concerning the facts of nicotine use, but there are dissenters to the view that those facts qualify nicotine for the status of an addictive drug. Some relevant facts are: the very large majority of smokers do not go a single day without a cigarette for a period of years or even decades; abstinence from smoking is accompanied in many cases by powerful cravings and those smokers in a position to compare cigarette cravings with other drug cravings put them on a similar footing; when a smoker makes a determined attempt to stop smoking for good he or she has a less than 5% chance of still being off cigarettes after one year; abstinence from cigarettes is often accompanied by severe mood disturbance; temporary nicotine substition (albeit at a lower level and with slower absorption than from cigarettes) significantly reduces withdrawal discomfort and increases the chances of continued abstinence. Against the free choice argument, likelihood failure of an attempt to give up smoking is not related to previously reported pleasure in smoking, nor to other perceived benefits, but is related to prior nicotine intake and cravings.

SCIENCE AND COMMON-SENSE SUPPORT THE VIEW THAT NICOTINE IS NOT ADDICTIVE.

John H. Robinson and Walter S. Pritchard. Psychophysiology Laboratory, Bowman Gray Technical Center, R.J. Reynolds Tobacco Company, Winston-Salem, NC 27102.

Recent events have publicly highlighted the major conclusion of the 1988 U.S. Surgeon General's report, namely, that nicotine in cigarette smoke is an 'addictive' drug in the same sense as drugs such as heroin and cocaine. The logic used to reach this conclusion along with the data supporting that logic have been questioned, most recently by Robinson & Pritchard (*Psychopharmacology*, 108, 1992, 397-407), where we argue that smoking is more accurately labeled as habit than addiction. Since there is no universally accepted definition of addiction, this position has generated extensive discussion and reaction (see West, *Psychopharmacology*, 108, 1992, 408-410, Hughes, *Psychopharmacology*, 113, 1993, 282-283; Stolerman and Jarvis, *Psychopharmacology*, in press). In contrast to the Surgeon General's report, we feel that the data support the position that the physiologic, pharmacologic and behavioral effects of nicotine absorbed from cigarette smoke are fundamentally different than the effects of addicting drugs such as heroin or cocaine, and more similar to the effects of caffeine absorbed from coffee. It will be argued that a key component of any meaningful definition of addiction is the concept of intoxication, a hallmark of several expert definitions of addiction. Labeling smoking and coffee drinking as habits as opposed to addictions emphasizes important distinctions between nicotine and caffeine and 'hard' drugs, and is compatible with reports that some people may have difficulty in giving up their habit or modifying their behavior. Compulsive use, psychoactivity and the reinforcing effects of nicotine, caffeine and addicting drugs will be discussed to support the argument that people smoke not because they are addicted, but because smokers find smoking a pleasurable activity with certain emotional and cognitive benefits.

INDIVIDUAL DIFFERENCES IN TOBACCO USE MAY BE RELATED TO GENETICALLY-DETERMINED DIFFERENCES IN RESPONSES TO NICOTINE. Allan C. Collins, Scott F. Robinson and Michael Marks. Institute for Behavioral Genetics, University of Colorado, Boulder, Colorado 80309, USA

Individuals differ in use and response to tobacco and twin studies suggest genetic factors may regulate individual differences. Since sons of alcoholics have reduced sensitivity to ethanol, which may predispose them to alcoholism, we determined whether inbred mouse strains differ in sensitivity to nicotine. Dose-response curves were constructed using 19 inbred mouse strains for seven different responses to nicotine. Isoeffective doses varied by approximately 3-5 fold depending on the response measured. Some of this variability correlated with differences in the number of brain nicotinic receptors; more sensitive strains have greater numbers of receptors. The development of tolerance to nicotine has been measured in seven different mouse stocks by infusing chronically with nicotine (0-6 mg/kg/hr) for 7 days. The strains varied in the minimum infusion dose that evoked detectable tolerance. Those strains that were most sensitive to a first dose of nicotine developed tolerance at the lowest chronic treatment doses. Nicotine preference using two-bottle choice methodologies has also been determined in six of these strains. Wide variance in preference was observed. The rank order for preference closely matched the rank order for first dose sensitivity. Thus, three potential components of the dependence process seem to be co-segregating, suggesting that individual differences in tobacco use by humans are related to genetically determined differences in sensitivity to acute and chronic responses to nicotine. Supported by DA-03194, DA-00197 and AA-06391.

Session 6: Nicotine and human diseases

S 38   RELATIONSHIP BETWEEN SMOKING, NICOTINE AND ULCERATIVE COLITIS. GAO, Thomas and J. Rhodes, Department of Gastroenterology, University Hospital of Wales, Heath Park, Cardiff CF4 4XW, Wales, U.K.

Ulcerative colitis is a disease of lifelong non-smokers and ex-smokers. The initial observation was made almost by chance in 1981 and showed that only 8 percent of 230 patients with colitis were current smokers compared with 44 percent of matched controls. Since then, many observations of patient populations both hospital and community based, with controls, have confirmed the finding. A meta-analysis on selected studies showed the relationship to be remarkably consistent in terms of the direction of findings; a comparison of former smokers with life-long non smokers showed a significantly greater risk among former smokers. Patients with colitis who are ex-smokers, usually develop their disease within a few years of cessation of smoking. There is little reliable data on the effect of smoking on activity of disease because of the inherent difficulties in such studies. However, half of a group of 30 patients who were intermittent smokers noted improvement of symptoms over six weeks while smoking twenty cigarettes daily. In attempts to identify the reason for benefit from smoking, nicotine has been tested formally using transdermal nicotine patches in a controlled trial involving 77 patients with active colitis; those given nicotine benefited significantly more than controls. Epidemiological data strongly supports the association of non-smoking with ulcerative colitis. Nicotine is probably the active ingredient responsible for this effect and may be of therapeutic value in active disease or prevention of relapse. Mechanisms responsible for the association remain an enigma but may involve mucosal eicosanoids and intestinal mucus.

S 39   BENEFICIAL EFFECTS OF NICOTINE IN TOURETTE'S SYNDROME

Paul R. Sanberg and Archie A. Silver, Departments of Surgery and Psychiatry, University of South Florida College of Medicine, Tampa, Florida 34610

Tourette's syndrome (TS) is a complex disorder with both motor and verbal tics. While the drugs of choice for TS are dopamine receptor blockers, such as haloperidol, some patients demonstrate only marginal improvement. In addition, these drugs can lead to sedation, exacerbation of learning difficulties, and possible tardive dyskinesia. For these reasons, an agent that potentiates neuroleptics could allow for lower doses of neuroleptics and reduction in side effects. In animals, it has been shown that nicotine markedly potentiates the behavioral effects of haloperidol. In addition, nicotine gum given to TS patients who only showed a moderate response to haloperidol alone demonstrated a marked reduction in the intensity in tics and an increase in attention and concentration. The effect with nicotine gum lasted about an hour, however, bitter taste and gastrointestinal side effects often caused non-compliance. In an open-trial we have examined the effect of transdermal nicotine patch on the amelioration of symptoms in TS patients. Preliminary case results so far suggest that (1) transdermal nicotine patch, titrated to deliver 7 mgms of nicotine in 24 hours, appears to enhance the therapeutic effect of dopamine blockers in non-smoking adolescents whose TS symptoms have not been effectively controlled with dopamine blockers; (2) the absorption rate of nicotine appears to be very rapid with effect seen approximately 2 to 3 hours after application of the patch, and; (3) the therapeutic effect of a single transdermal nicotine patch seems to be persistent for a variable period of time (lasting from days to weeks) even after the patch is removed. This long-term effect from a single patch was unexpected. Supported in part by the Smokeless Tobacco Research Council and NINDS R01NS32067-01.

S 40   NICOTINE AND NEUROPSYCHIATRIC DISORDERS. J. Hughes. Dept. of Psychiatry, University of Vermont, Burlington, VT, USA 05405

Smoking is highly prevalent and cessation rare in patients with alcohol/drug abuse, attention-deficit disorder, depression and schizophrenia. Several preliminary findings in humans and nonhumans illustrate how nicotine via smoking could interact with psychiatric illness; e.g., nicotine and alcohol can produce cross-tolerance, there is a shared genetic predisposition to alcohol and to nicotine use, nicotine increases the discrimination and self-administration of alcohol, many of the symptoms of nicotine withdrawal are similar to those of attention-deficit disorder, smokers have low MAO levels, smoking decreases neuroleptic and antidepressant levels, nicotine increases neuroleptic-induced akathesia and perhaps tardive dyskinesia yet decrease parkinsonian tremor. Programmatic research into these areas could suggest neuropharmacological reasons why smoking is prevalent and cessation rare in psychiatric and alcohol/drug abuse patients.

NICOTINE, AUDITORY GATING, AND SCHIZOPHRENIA. R. Freedman, L.E. Adler, P. Bickford, V. Luntz-Leybman, K. Wear, L.J. Hoffer, J. Griffith, M. Waldo, H. Coon, M. Myles-Worsley, S. Leonard, W. Byerley, Depts. of Psychiatry and Pharmacology, University of Colorado and Denver VAMC, Denver, CO, USA 80262 and Dept. of Psychiatry, University of Utah, Salt Lake City, UT, USA 84132. S 41

Schizophrenics are often unable to focus their attention and to filter out extraneous noise. An underlying deficit in the inhibitory gating or habituation of a cerebral evoked response (P50 wave) to repeated auditory stimuli has been demonstrated. Animal models of similar evoked responses indicate a source in the hippocampus. Fimbria-fornix lesions in the animals block such habituation, but it is restored by nicotine administration. Pharmacological studies indicate that alpha-bungarotoxin-sensitive receptors are involved. Nicotine also normalizes inhibitory gating of the P50 evoked response in schizophrenics, as well as in their relatives who share the deficit. Further pharmacological studies in a human subject suggest that a low affinity nicotinic receptor is involved, as this effect is not blocked by mecamylamine. These effects of nicotine may partly explain the heavy smoking of many schizophrenics. Linkage studies in nine multi-affected pedigrees of schizophrenics, using the illness and the P50 evoked potential abnormality as alternative phenotypes, showed modestly positive LOD scores for the P50 phenotype, above the 95% confidence level expected for unlinked markers, which could be verified by flanking markers for one or more of the families in four chromosomal locations. LOD scores for schizophrenia were also positive, but lower than found for the P50 phenotype. One of these locations, 15q14, has been identified as the site of the alpha-7 nicotinic receptor gene. Further studies are needed to determine the alpha-7 gene's role in the genetic transmission of schizophrenia.

EPIDEMIOLOGY OF SMOKING AND PARKINSON'S DISEASE. John A. Baron, Department of Medicine, Dartmouth Medical School, Hanover, NH, 03756. S 42

An inverse association between cigarette smoking and Parkinson's disease (PD) was first noted almost 30 years ago; subsequently this finding has been confirmed in over 25 investigations. The relevant data are derived from studies of various design, performed in different clinical or population settings in a variety of geographic locations. Although many of the studies have limitations, the data are relatively consistent: men or women who have ever smoked cigarettes have about half to two thirds the risk of never smokers. In several studies, heavy smokers or longer-term smokers had a lower risk than those who smoked fewer cigarettes, or who quit smoking altogether. The inverse association persists even if smoking early in life (before the onset of symptoms) is considered. Selective mortality of PD cases who smoke is not an explanation for the observed relationship, since follow-up studies have also found it, as have investigations of incident cases. Several investigators have found PD symptoms to start at an earlier age in smokers than in non-smokers. This pattern is consistent with a protective impact of smoking, since in these studies smokers have been younger than non-smokers. It is conceivable that some factors related both to PD and to non-smoking might underlie the inverse relationship. Such factors must be strongly related to both PD and smoking in order to provide an explanation for the low PD risk in smokers. Currently no known factor, including a "pre-Parkinsonian personality" seems to have these characteristics. An inverse relationship between smoking and drug-induced Parkinsonism, and case reports of symptomatic improvement in PD after smoking, provide further support for a genuine protective effect of cigarette smoking on Parkinson's disease.

NICOTINE AND ANIMAL MODELS OF PARKINSON'S DISEASE. A.M. Janson, *A. Møller, P.B. Hedlund, G. von Euler and K. Fuxe. Dept. of Neuroscience, Karolinska Institute, S-171 77 Stockholm, Sweden, *NeuroSearch, Glostrup, Denmark. S 43

In view of the negative correlation between smoking and Parkinson's Disease seen in epidemiologic studies, we have investigated whether chronic continuous (-)nicotine treatment exerts protective actions in lesioned nigrostriatal dopamine systems using two different animal models. In the first model nicotine was delivered for two weeks by subcutaneously implanted osmotic pumps to male Sprague-Dawley rats with a partial unilateral mesodiencephalic lesion. Nicotine caused a significant counteraction of the lesion-induced reduction in total number of nigral tyrosine-hydroxylase-like immunoreactive neurons counterstained with cresyl violet compared with saline treated control animals. Combining immunocytochemical and stereological methods, unbiased estimates of the total number of both neuronal and non-neuronal cells in the entire substantia nigra and their respective mean volumes were studied. Chronic nicotine treatment also counteracted the lesion induced upregulation of the high-affinity binding sites of the striatal dopamine D_2 receptors analyzed with [3H]N-propylnorapomorphine. Furthermore, nicotine induced a disappearance of substance P-like immunoreactivity in the substantia nigra ipsilateral to the lesion compared with saline-treated control animals. The results indicate that chronic nicotine treatment has protective actions on partially lesioned nigrostriatal dopamine systems. The mechanism underlying the protective effects could be explained by a functional desensitization of the nicotinic cholinoceptors located on dopamine nerve cells. Together with a decreased excitation of the dopamine cells by substance P this would lead to reduced energy demands to maintain ion homeostasis and reduced Ca^{2+} influx into the cell leading to an increased dopamine nerve cell survival. In a different lesion model, nicotine was given via osmotic minipumps to male C57B1/6 mice treated with 1-methyl-4-phenyl-1,2,3,6-tetrahydropyridine. Depending on the dose-schedule, nicotine either attenuated or enhanced the MPTP-induced neurotoxicity. The latter could be explained by an action of MPTP on the nicotinic cholinoceptors preventing functional desensitization to take place.

S 44   MEMORY ENHANCING EFFECTS OF NICOTINE. Edward D. Levin and Diane Torry, Depts. of Psychiatry and Pharmacology, Duke University Medical Center, Durham, NC 27710, USA and Dept. of Biochemistry, University of Bath, BA2 7AY, Bath, UK.

Nicotine has been found in a variety of studies to improve memory performance. The neural mechanisms and psychological processes underlying this effect are currently under investigation. Other studies have not found nicotine-induced memory improvement and some have documented nicotine-induced deficits. These discrepant findings present the opportunity for uncovering the nature of nicotine effects on memory function and the mechanisms underlying it. In our laboratory, a very reproducible finding has been that chronic infusion of 12 mg/kg/day of nicotine significantly improves choice accuracy in a win-shift version of the radial-arm maze ($p<0.0001$). We have seen this effect in seven studies including 60 controls and 63 nicotine-treated rats. The effect is robust and is still seen after knifecut lesions of the fimbria-fornix or the medial basalocortical projection. Nicotine attenuates the adverse effects of these lesions on memory performance. We have not found concurrent administration of the D_2 agonist quinpirole or the D_2 antagonist raclopride to significantly affect the nicotine-induced improvement. It is eliminated by either acute or chronic administration of the nicotinic antagonist mecamylamine. In a T-maze spatial alternation task, we did not find a significant effect of the same dose of chronic nicotine on choice accuracy. The rats not given nicotine (N=11) averaged 83.8±2.2 percent correct, while those given nicotine (N=10) averaged 80.4±1.5 percent correct. Mecamylamine improved performance at the 10 second delay ($p<0.05$) during week 1, but later caused a deficit at the 0 delay during week 4 ($p<0.05$). The effects of nicotine appear to be dependent upon the specific nature of the memory task. Nicotine-induced increased proactive interference may have attenuated its effect on memory performance in the T-maze. Paradoxical mecamylamine-induced improvements suggest that the antagonistic effects of nicotine may be important for its effects on memory. (Research supported by the Council for Tobacco Research-USA and the Alzheimer's Association)

S 45   POSSIBLE MECHANISMS UNDERLYING BENEFICIAL EFFECTS OF NICOTINE ON COGNITIVE FUNCTION.   M.H. Joseph, G. Grigoryan, H. Hodges and J.A. Gray
MRC Behavioural Neurochemistry Group and Department of Psychology,
Institute of Psychiatry, Denmark Hill, London SE5 8AF, UK

Nicotine improves performance in cognitive tasks both in patients with senile dementia of Alzheimer type (AD), and in animal models of the disorder, induced either by neurotoxic lesions to the cell bodies of the forebrain cholinergic projections, or by chronic alcohol administration. The nature of the improvement suggests a generalised action on "vigilance", rather than a specific action on learning or memory. While these actions may be mediated through denervated cholinergic synapses in the forebrain, nicotine acts at many sites in the brain. In animals it releases catecholamines; increased dopamine release in the mesolimbic system appears to underlie the effects of nicotine on latent inhibition, an attentional paradigm, while increased noradrenaline release in the coerulo-hippocampal projection may result in increased resistance to extinction. Recent evidence suggests that both of these effects of nicotine are mediated at the cell bodies, rather than at the terminals of the respective projections. If nicotine releases catecholamines also in human brain, for which there is only limited evidence, this action could underlie its above effects in AD. However, our continuing animal experiments (see Gray, Mitchell, Joseph *et al.*, Drug Dev. Res. **31** 3-17, 1994) indicate that the effects of nicotine on "vigilance" in animals are independent of its effects on noradrenaline release in the hippocampus, and at least partly independent of its effects on dopamine release in the n. accumbens.

MHJ is a member of the MRC (UK) External staff, HH is a Wellcome senior lecturer; we also thank CTR (USA) for financial support.

S 46   NICOTINIC MODULATION OF COGNITIVE FUNCTIONING IN HUMANS
Paul A. Newhouse, Alexandra Potter, Melissa Piasecki, Jennifer Geelmuyden, June Corwin, Robert Lenox
Department of Psychiatry, University of Vermont College of Medicine, 1 South Prospect St., Burlington, VT 05401.

The loss of central nicotinic receptors is a neurochemical hallmark of several degenerative brain disorders, notably Alzheimer's (AD) and Parkinson's disease (PD). Evidence from our laboratory suggests that there appear to be both age- and disease-related alterations in the sensitivity of certain cognitive domains (new learning, attention, retrieval) to central nicotinic agents. The central and peripheral nicotinic antagonist mecamylamine was administered to healthy young and elderly, Alzheimer's (AD), and Parkinson's disease (PD) subjects in doses of 5, 10, and 20 mg in a placebo-controlled, double-blind study. In all groups, mecamylamine produced interference with new learning, while not affecting long-term retrieval. On the recognition memory task, there was an age-related shift in response bias with the elderly subjects becoming more liberal with increasing dose. Reaction time measures suggested a dose-related slowing of RT on several tasks. Behavioral effects were minimal and physiologic measures were consistent with dose-related ganglionic blockade. There was progressive increases in sensitivity to nicotinic blockade with increasing age and disease on several cognitive tasks; elderly normal and AD subjects showed enhanced sensitivity to mecamylamine while PD subjects did not. These results indicate that acute blockade of nicotinic receptor function can produce measurable and significant cognitive impairment similar in nature to deficits seen in dementing illnesses and that there is an age-related enhancement of sensitivity to nicotinic blockade. These data and others indicate that symptoms of impaired acquisition of information and short-term storage, attention, visual perception, and speed may reflect nicotinic lesions. These studies support the importance of intact CNS nicotinic mechanisms for normal cognitive functioning. Further, these results and results of other studies suggest that nicotinic augmentation may be a useful strategy for cognitive enhancement in certain dementias and other conditions with cognitive impairments. Supported by NIMH R29-46225

Poster Presentations

Section 1: Receptor localization

DISTRIBUTION OF NICOTINIC ACETYLCHOLINE RECEPTOR SUBUNIT IMMUNOREAC-
TIVITIES ON THE SURFACE OF CHICK CILIARY GANGLION NEURONS. P.B Sargent and
H.L. Wilson. Depts. of Stomatology and Physiology and the Neurosciences Graduate
Program, UCSF, San Francisco, CA 94143.

The distribution of acetylcholine receptor (AChR) subunit-like immunoreactivities (LI) on the surface of chick ciliary ganglion neurons was examined by indirect immunofluorescence and a laser scanning confocal microscope. AChRs were immunolabeled with rat or mouse anti-AChR mAbs provided by Dr. Jon Lindstrom (UPenn) and Dr. Darwin Berg (UCSD). Synaptic sites were visualized with a mouse anti-SV2 mAb provided by Dr. Steve Carlson (UWash). Bound primary mAbs were visualized with secondary Abs conjugated with cyanine 3.18 or cyanine 5.18. Intracellular staining was noted in all neurons with antibodies specific for each of the subunits known to be expressed by the ganglion as a whole ($\alpha3$, $\alpha5$, $\alpha7$, $\beta2$, $\beta4$). At the neuronal surface $\alpha5$-LI was distributed both at synaptic and extrasynaptic sites, while $\alpha7$-LI was distributed exclusively at extrasynaptic sites. These results are consistent with previous electron microscopic observations of Jacob and Berg (J. Neurosci. 3: 260-271, 1983) and Jacob et al. (PNAS 81: 3223-3227, 1984). A5-$\alpha7$ dual labeling showed some areas of co-localization and others areas where each subunit LI was present in the absence of the other. Rhodamine-α-bungarotoxin co-localized precisely with anti-$\alpha7$ mAbs, in keeping with its documented ability to block $\alpha7$ function. In initial trials none of the available mAbs specific for $\alpha3$, $\beta2$, or $\beta4$ produced cell-surface staining. (Supported by NIH NS 24207.)

EFFECTS OF DENERVATION UPON NICOTINIC ACETYLCHOLINE RECEPTOR
CLUSTERS IN AUTONOMIC NEURONS AS DETERMINED BY QUANTITATIVE LASER
SCANNING CONFOCAL MICROSCOPY. P.B Sargent and H.L. Wilson. Depts. of
Stomatology and Physiology and the Neurosciences Graduate Program, UCSF, San
Francisco, CA 94143.

We previously showed that denervation altered the properties of nicotinic acetylcholine receptor (AChR) clusters on the surface of frog cardiac ganglion neurons (Neuron 1: 877-886, 1988). We have extended this analysis with the aid of a laser scanning confocal microscope, which was used to optically section and reconstruct neurons labeled both for AChR clusters and synaptic boutons. AChR clusters were labeled with a rat anti-electric organ AChR antibody (gift of Dr. Jon Lindstrom, UPenn) and cyanine 3.18-labeled goat anti-rat IgG, while synaptic boutons were labeled with a mouse anti-SV2 antibody (gift of Dr. Steve Carlson, UWash) and cyanine 5.18-labeled goat anti-mouse IgG. Denervation leads to a profound change in the distribution of AChR clusters, which are spatially clustered normally, often with several clusters associated with a single bouton, but which are spatially dispersed following denervation. Denervation also leads to a reduction in the relative AChR number per cluster, measured as the product of cluster size and average pixel value; this change is evident as soon as 3 days following surgery and is sustained for at least 6 weeks. Denervation also results in a transient change in the relative AChR number per cell, which is significantly reduced 18-20 days following denervation but which returns to normal by 40-42 days following denervation. (Supported by NIH NS 24207.)

CELLULAR DIVERSITY IN THE EXPRESSION OF NICOTINIC ACETYLCHOLINE
RECEPTOR SUBUNITS IN THE CHICK CENTRAL NERVOUS SYSTEM. E.M. Ullian and
P.B. Sargent. Neurosciences Graduate Program and the Depts. of Stomatology and
Physiology, UCSF, San Francisco, California 94143.

The heterogeneity of nicotinic acetylcholine receptor (AChR) expression in the chick lateral spiriform nucleus (SpL) was assessed with monoclonal antibodies (mAbs) specific for various AChR subunits [provided by Dr. Jon Lindstrom (UPenn) and Dr. Darwin Berg (UCSD)]. Bound primary mAbs were visualized with secondary Abs conjugated with cyanine 3.18 or cyanine 5.18 and with a laser scanning confocal microscope. The SpL stained positively with antibodies specific for the $\alpha2$, $\alpha5$, $\alpha7$, $\alpha8$, and $\beta2$ subunits. Cell classes in the SpL were determined by comparing the pattern of pairwise staining of subunit-specific mAbs. Approximately 90% of the neurons in the SpL contained both $\alpha5$ and $\beta2$ AChR subunit-like immunoreactivity (LI), and few, if any, cells were immunoreactive for only one of these two subunits. Fewer cells, approximately 70%, contained $\alpha2$-LI, while roughly 30% of the cells in the SpL contained $\alpha7$-LI. All $\alpha2$-LI containing cells were also $\alpha5$- and $\beta2$-LI positive, and approximately two-thirds of the $\alpha7$-LI positive cells were also $\alpha5$-LI positive and $\beta2$-LI positive. These results suggest that there are several populations of neurons within the SpL, based on their expression of AChR subunits (see Table) and reveal a significant degree of diversity of AChR subunit expression at the cellular level within the CNS. (Supported by NIH NS 24207 and GM 07449.)

Class	$\alpha2$-LI	$\alpha5$-LI	$\alpha7$-LI	$\beta2$-LI
1	-	+	-	+
2	+	+	-	+
3	+	+	+	+
4	-	-	+	-

P 4

CELLULAR AND SUBCELLULAR VISUALIZATION OF THE ß2-SUBUNIT OF THE NICOTINIC ACETYLCHOLINE RECEPTOR IN THE MOUSE CEREBRAL CORTEX. <u>Reinhard Marks</u>[1], <u>John Lindstrom</u>[2], <u>Hannsjörg Schröder</u>[1]. [1]Department of Anatomy, University of Cologne, Cologne, Germany, [2]Institute for Neurological Sciences, University of Pennsylvania Medical School, Philadelphia, PA, USA.

The cholinergic innervation of the rodent cerebral cortex and the hippocampus arising from basal forebrain nuclei plays an important role in the modulation of signal transduction in the telencephalon. Furthermore, there is evidence for the involvement of central nervous acetylcholine and of the nicotinic acetylcholine receptor (nAChR) in pathological states, e.g. neurodegenerative diseases. The mouse may serve as a useful model to mimic certain aspects of such diseases. However, so far little is known about the location of cholinergic synapses and the cellular expression of nAChR in the mouse telencephalon. In order to achieve the visualization of the nAChR on a light and electron microscopic level we used the monoclonal antibody (mAb) 270 (Whiting et al., *Proc Natl Acad Sci, USA*, 84:595, 1987), which is directed against the structural ß2 - subunit of the nAChR. MAb 270 binding sites were visualized using a biotin streptavidinperoxidase detection pattern of immunolabeled cells in the cerebral cortex was characterized by large pyramidal perikarya in layers II/III and V and their basal and apical dendrites, regardless of the cytoarchitectonic features of distinct areas. Pyramidal apical dendrites could be traced up to layer I/II where they split and formed a dendritic net in layer I. In addition, a considerable number of smaller, labeled neurons could be found in layer VI of the primary somatosensory cortex and in layer IV of the primary visual area. The hippocampus showed positive staining of apical dendrites and perikarya of CA1 and CA3 pyramidal neurons and a densely packed band of immunoreactive neurons in the granular layer of the dentate gyrus. On the electron microscopic level the immunoprecipitate was associated with dendritic microtubuli and could also be observed in postsynaptic thickenings opposite to vesicle-containing synaptic terminales. These findings correlate well with the immunocytochemical demonstration of nAChR subunits (α,ß) in rat and human brain. Furthermore, our results are comparable with the expression of ß2-subunit mRNA and the distribution of cholinergic fibres as revealed by in situ hybridisation and ChAT - immunocytochemistry. ß2-subunits immunohistochemistry will be a valuable tool for the assessment of central nervous nAChR expression. Supported by the Deutsche Forschungsgemeinschaft (R.M., HJ.S.) and by grants from NIH, MAD, Council for Tobacco Research and Council for Smokeless Tobacco Research (J.L.).

P 5

IMMUNOCHARACTERIZATION OF THE HUMAN ALPHA 7 NEURONAL NICOTINIC ACETYLCHOLINE RECEPTOR SUBUNIT. <u>M. Piattoni-Kaplan, D. Donnelly-Roberts, J. Pauly*, D. Hill*, J.B. Pan, S.P. Arneric, and J.P. Sullivan.</u> Neuroscience Research, D-47W, Abbott Laboratories, Abbott Park, Illinois, U.S.A., 60064-3500 and *Department of Pharmacology, Medical College of Georgia, Augusta, Georgia, U.S.A., 30912.

Accumulating preclinical and clinical data implicate an important role for the α7 subtype of the neuronal nicotinic acetylcholine receptor in a number of neurological disorders. The human α7 (huα7) subunit, expressed as a functional, homomeric channel in *Xenopus* oocytes, has similar, yet discretely different pharmacological properties when compared to its rat and chick homologs (X. Peng *et al.*, *Mol. Pharm.*, in press.). In the present study, the immunocharacterization of the huα7 subtype is described. Polyclonal antibodies (huα7-Ab) were generated using a multiple antigenic peptide (MAP) corresponding to the amino acid sequence 320-329 present in the cytoplasmic region of the mature human and rat α7 protein sequences and was purified on a peptide affinity column. Immunodetection of huα7 was performed using the human neuroblastoma cell line IMR32. These cells express a high density of ^{125}I labeled α-bungarotoxin binding sites (~150 fmol/mg of protein). Western analysis of protein from IMR32 cells and oocytes injected with huα7 cRNA detected a band of 59kD that was eliminated by preincubation with peptide. These data suggest the presence of huα7 protein in IMR32 cells. Pilot immunohistochemical analysis of rat brain sections demonstrated specific labeling with the antibody in the mammillary nuclear complex, interpeduncular nucleus, superior colliculus, dentate gyrus, ventrolateral geniculate nucleus, red nucleus, and the nucleus of Edinger-Westphal, confirming the expected cross-reactivity of the huα7-Ab with the rat α7 subunit. The staining correlates well with localization of α7 mRNA and α-bungarotoxin binding sites. These data suggest the huα7-Ab will be a useful probe for the localization of α7 in human brain. Ultimately, such experiments may reveal new insights into the pathology, progression and treatment of Alzheimer's disease and schizophrenia.

P 6

LOCALIZATION OF [^{3}H]CYTISINE NICOTINIC BINDING SITES IN NORMAL AND PATHOLOGICAL HUMAN BRAIN USING *IN VITRO* RECEPTOR AUTORADIOGRAPHY. <u>I. Aubert, D. Cécyre, S. Gauthier and R. Quirion.</u> Douglas Hospital Research Center, Departments of Psychiatry, Neurology & Neurosurgery McGill University, and McGill Center for Studies in Aging, Montreal, Quebec, Canada H4H 1R3.

In the present study, we investigated the status of [^{3}H]cytisine/nicotinic binding sites, using *in vitro* receptor autoradiography, in postmortem human brains from Alzheimer's, parkinsonian and control patients. [^{3}H]Cytisine generated high quality autoradiograms in comparison to previously used radioligands for α-bungarotoxin insensitive nicotinic receptors. Highest levels of [^{3}H]cytisine sites were found in the lateral geniculate body and various thalamic nuclei. The substantia nigra, middle layers of frontal, temporal and entorhinal cortices, the subiculum, and the molecular layer of the cerebellum contained relatively high densities of [^{3}H]cytisine binding. Moderate amounts of [^{3}H]cytisine sites were observed in the striatum, superficial and deep cortical laminae, outer two-thirds of the molecular layer of the dentate gyrus, and granular layer of the cerebellum. Subfields of the Ammon's horn of the hippocampus demonstrated low levels of [^{3}H]cytisine binding; the CA1 region contained less [^{3}H]cytisine sites compared to the CA2 and CA3 subfields. The density of [^{3}H]cytisine binding was very low in the inner third of the molecular layer of the dentate gyrus, granular layer of the dentate gyrus, and the globus pallidus. Hence, [^{3}H]cytisine should prove to be a most useful probe for detailed investigation of the status of nicotinic receptors, probably of the α4β2 subtype, in human neurodegenerative disorders. Supported by the Alzheimer Society possibly of Canada and MRCC.

CLONING AND FUNCTIONAL EXPRESSION OF ALPHA 9: A NOVEL ACETYLCHOLINE-GATED ION CHANNEL. AB Elgoyhen, D Johnson, J Boulter, D Vetter & SF Heinemann. Molecular Neurobiology Laboratory, The Salk Institute for Biological Studies, La Jolla, California.

Molecular cloning studies have demonstrated structural and functional diversity in neuronal nicotinic acetylcholine receptors. To date, seven α subunits ($\alpha2$ to $\alpha8$) and three β subunits ($\beta2$ to $\beta4$) have been characterised in the nervous system of vertebrates. We now report the identification and functional characterization of a new member of this family of receptor subunit genes: $\alpha9$. A full-length cDNA clone encoding a protein of 480 amino acids was isolated from an olfactory epithelium library. The deduced amino acid sequence reveals that $\alpha9$ has significant homology with other nicotinic acetylcholine receptor α subunits. In situ hybridization studies using sagittal sections of rat embryos show that this subunit is expressed in the olfactory epithelium and in the pars tuberalis of the pituitary. In vitro transcribed $\alpha9$ cRNA injected into Xenopus laevis oocytes results in the formation of a homomeric, cationic channel which is activated by acetylcholine with an EC_{50} of $10\mu M$. While acetylcholine is a full agonist, both the nicotinic agonist DMPP and the muscarinic agonist oxotremorine-M are weak partial agonists. The receptor is not activated either by nicotine or by muscarine. However, both nicotine and muscarine behave as competitive antagonists with an IC_{50} at $10\mu M$ acetylcholine of 45 μM and 84 μM, respectively. Classical nicotinic (d-tubocurarine, β-erythroidine) and muscarinic (atropine) antagonists also block the responses elicited by acetylcholine. Both α-bungarotoxin (100nM) and κ-bungarotoxin (100nM) block the recombinant receptor in a reversible manner. In summary, these heterologous expression studies show that the receptor-channel complex formed with $\alpha9$ has a mixed nicotinic-muscarinic pharmacology. In vivo receptors with similar properties have been reported in cochlear hair cells of chick and rat (Fuchs and Murrow, *Proc. R. Soc. Lond. B.* 248: 35, 1992; Housley and Ashmore, *Proc. R. Soc. Lond. B.* 244: 161, 1991). PCR experiments using specific primers for $\alpha9$ and cDNA reverse transcribed from rat cochlear total RNA show that this subunit gene is transcribed in this tissue.

SEARCHING FOR AN ACETYLCHOLINE-GATED CHLORIDE CHANNEL: ANALYSIS OF CLONED LEECH NICOTINIC GENES. R. Allen, M. Hartley and S. Heinemann. Molecular Neurobiology Laboratory, The Salk Institute for Biological Studies, La Jolla, CA 92037 USA.

To date, the ion channels formed by the expression of cloned nicotinic acetylcholine receptor (nAChR) genes are cation permeable. However, chloride channels activated by ACh and nicotine have been observed in neurons of several invertebrate species, such as Retzius and HN cells in leech, H cells in snail and Aplysia ganglion cells. The aim of this project is to characterize the molecular nature of this unique response to ACh. In the leech, Retzius neurons hyperpolarize or depolarize in response to ACh depending on the segmental location of their ganglia. Several lines of evidence suggest that the different responses are mediated by novel nAChRs. A cloning strategy has been devised assuming that these receptors will share domains conserved in other nAChR genes. Total RNA prepared from adult leech nerve cord or muscle was reverse transcribed, and then degenerate oligonucleotide primer sets based on conserved regions were used to amplify the resulting cDNAs by PCR. Products of ~500 base pairs (bp) in length corresponding to the nAChR extracellular domain extending through transmembrane region (TM)3 were isolated, subcloned and sequenced to determine whether they contain 1) canonical, paired Cys residues found in other cloned α subunits, and 2) a unique TM2 structure, the putative ion channel. Seven different PCR products have been identified, including two coding for α-like subunits based upon paired Cys residues. The cDNAs coding for non-α-like subunits range in length from 447 bp (clones L7, L14, L28 and L49) to 513 bp (clone L13). The α-like clones are 471 bp (Lαn) and 474 bp (Lαm). L7, L14 and Lαm were isolated from muscle RNA; L13, Lαn from nerve cord RNA, and L28 and L49 were isolated from both RNAs. All share extensive sequence homology with cloned nicotinic genes. The α's share 70% identity at the level of the DNA, and 78% identity and 87% similarity at the protein level. Northern blot analysis reveals specific hybridization of probes prepared from Lαn to nerve cord RNA, and Lαm to muscle RNA. Using Lαm as a probe at high stringency, a full length clone has been isolated from a leech embryo cDNA library. Functional analysis of this clone by expression in Xenopus oocytes is currently in progress.

HETEROLOGOUS EXPRESSION OF EPITOPE-TAGGED NEURONAL NICOTINIC AND 5HT3A RECEPTOR SUBUNITS. Jishnu Mukerji, Éric Dumont and Philippe Séguéla. Neurobiology Unit, Montreal Neurological Institute, McGill University, Montreal, Quebec, Canada H3A 2B4.

The discovery of numerous nicotinic acetylcholine receptor subunits ($\alpha2$-8, $\beta2$-4) expressed in the vertebrate nervous system leads to the issue of stoechiometry of neuronal acetylcholine-gated channels. It is assumed that the large number of theoretical combinations of alphas and betas in a pentamer is restricted both by structural constraints embedded in the primary sequence of the subunits and by transcriptional regulation. To address this issue of the composition of neuronal acetylcholine-gated channels and to avoid the problems of immunological cross-reactivity between homologous gene products, we developped a subunit-specific in vitro strategy of epitope-tagging using the mutated C-terminal domain of the proteins as an anchor for addition of hydrophilic peptide antigens. The successful expression of functional homo-oligomeric tagged α-bungarotoxin-sensitive $\alpha7$ channels and curare-sensitive 5HT3A serotonin-gated channels in Xenopus oocytes validated the structural model in which the domain of these ligand-gated channels located 3' of the putative TMD4 does not play a role in assembly or in activation of the receptor complex. Using immunoprecipitation from membranes of Xenopus oocytes and from transiently transfected mammalian cells, the application of this labeling method to naturally co-expressed components of hetero-oligomeric receptors will allow us to precise the set of possible subtypes with unique functional properties by direct identification of co-assembled subunits. (Supported by MRC, MT-12549)

P 10

NORTHERN BLOT ANALYSIS DEMONSTRATES THE PRESENCE OF THREE DIFFERENT TRANSCRIPTS OF NEURONAL NICOTINIC ACETYLCHOLINE RECEPTOR α4 GENE IN RAT BRAIN. Z. J. Yu, D. G. Morgan and L. Wecker. Department of Pharmacology and Therapeutics, USF College of Medicine, Tampa, FL 33612.

The initial report describing the neuronal nicotinic receptor α4 RNA identified 3 different sized messages believed to represent alternative splicing of a single transcript (Cell 48: 965, 1987). Although it has been shown that α4 is the most widely expressed α subunit RNA in rat brain, the abundance ratios of the transcripts in brain regions have not been described. In this study, purified poly(A)+ mRNA was size-fractionated and blotted to nylon membranes. Signals of α4 mRNA were detected by Northern hybridization with radiolabelled plasmid DNA for α4-1 or Hinfl fragment of α4-1. Quantification was performed by densitometry of X-ray film. Three transcripts homologous to α4-1 cDNA were detected in all brain regions tested corresponding to approximately 2.6 kb, 4.6 kb and 6.0 kb. In all brain regions investigated, the 6.0 kb transcript was the least abundant. In the midbrain, basal forebrain and striatum, the 4.6 kb transcript was most abundant and was 2.6, 1.6 and 1.6 times the abundance of the 2.6 kb transcript, respectively. In contrast, in hippocampus, thalamus/hypothalamus, midcerebral cortex and frontal cortex, the 2.6 kb transcript was most abundant; the ratios of the 2.6:4.6:6.0 kb transcripts were 1.0:0.71:0.33, 1.0:0.65:0.34, 1.0:0.47:0.09, and 1.0:0.41:0.14, respectively. When hybridization stringency was increased in an attempt to eliminate hybridization signals with partial complementation to the probe, all 3 bands were diminished in parallel, indicating a similar degree of homology to the α4-1 probe. This study demonstrates the presence of 3 alternatively processed transcripts of the neuronal nicotinic receptor α4 gene, and indicates that the relative abundance of these transcripts differs in different brain regions. (DGM is an Established Investigator of the American Heart Association. Supported by grant #0411 from the STRC, Inc.)

P 11

TRANSCRIPTIONAL REGULATION OF HUMAN α3 NICOTINIC SUBUNIT
D. Fornasari, E. Battaglioli and F. Clementi. CNR Center of Cytopharmacology, Department of Medical Pharmacology, University of Milan, Milan, Italy.

Neuronal nicotinic receptors are expressed in the central nervous system (CNS), autonomic ganglia and adrenal medulla. Different subunits (α2-α8; β2-β4)) have been identified that, appropriately assembled, can generate several different receptor isoforms, with distinct localization and pharmacological profile. The human gene encoding α3 is located on the chromosome 15, between β4 and α5. These subunits are co-expressed, and probably assembled in the same receptor molecule, in peripheral neurons, but not in the CNS, where they can be independently expressed in the different areas.
Northern analysis showed that the transcript for α3 is expressed in human neuroblastoma cell lines, such as IMR 32 or SK-N-BE, but not in nonneuronal cells, such as HeLa, TE671 or A431, even when RT-PCR was employed.
In order to understand the transcriptional mechanisms responsible for the expression of human α3, we have isolated the 5' flanking sequences of the gene and tested their ability to drive the expression of the luciferase reporter gene, by transient and stable transfections of neuronal and nonneuronal cell lines. A 1 Kb fragment, corresponding to the sequences immediately upstream of the start codon, is able to increase the luciferase activity by 20 times over the background, indicating that it should contain the core promoter and some of the cis-acting elements responsible for the expression of α3. This transcriptional activity is negatively regulated by upstream sequences, contained in a region of 2.3 Kb, that decrease the expression of the reporter gene by 5 times, a phenomenon consistent with the existence of a silencer. Neither the promoter nor the silencer displayed any preferential activity in neuronal or nonneuronal cell lines, suggesting that the cis-acting elements involved in the tissue-specificity might reside in other part of the gene. However, run-off experiments on intact nuclei showed that the transcription of α3 is efficiently initiated even in nonneuronal cells, where the mature transcript is undetectable. This finding points out the role of post-transcriptional mechanisms in the tissue-specific expression of human α3.

P 12

REGIONS OF β2 AND β4 THAT AFFECT THE ACH DOSE-RESPONSE RELATIONS OF NEURONAL NICOTINIC RECEPTORS. B. Cohen, A. Figl, M. W. Quick, C. Labarca, N. Davidson, and H. A. Lester. Division of Biology, California Institute of Technology, Pasadena, CA 91125, USA

We constructed chimeras and mutations of the rat β2 and β4 neuronal nicotinic subunits to locate the regions that contribute to differences between the acetylcholine (ACh) dose-response relationships of the α3β4 and α3β2 receptors. Expressed in Xenopus oocytes, the α3β2 receptor displays an EC_{50} for ACh ~twentyfold < the EC_{50} of the α3β4 receptor; and an apparent Hill slope (n_{app}) twofold less. Recovery of the EC_{50} and the n_{app} of the wild-type ACh dose-response relationships in a chimera requires an extensive region of β2 (β2:93-323) and β4 (β4:1-301); however, β4:1-109 is sufficient to increase the EC_{50} for ACh to a β4-like value and β2:1-120 is sufficient to reduce the EC_{50} for ACh to a β2-like value. Moreover, substituting just β2:103-120 for β4:105-122 is sufficient to endow the chimera with β2-like values for (a) the n_{app} for ACh, (b) the 30 μM cytisine to 30 μM ACh response ratio, and (c) the 100 μM tetramethylammonium to 30 μM ACh response ratio. Two β4/β2 residue substitutions at β2:Met101 and β2:Phe106 raise the EC_{50} of the chimeras from β2-like to β4-like values but mutations of these residues alone does not convert the characteristics of the ACh dose-response relationship to β4-like values. These results suggest that (a) residues β2:92-120 and β4:94-122 participate in an agonist binding site at the α/β interface and (b) the sequences between M1 and the M3-M4 loop are critical in restoring the n_{app}'s of the chimeric dose-response relationships for ACh to wild-type values. We suggest that these latter sequences include a domain of contact between the subunits that participates in the conformational change which opens the channel.

MAPPING DETERMINANTS OF COMPETITIVE ANTAGONIST SENSITIVITY ON NEURONAL NICOTINIC RECEPTOR SUBUNITS. C.W. Luetje, S.C. Harvey and F. Maddox. Department of Molecular and Cellular Pharmacology, University of Miami, Miami FL 33101.

Neuronal nicotinic acetylcholine receptors (nAChR) can be formed in *Xenopus* oocytes by injecting various combinations of cRNAs encoding two classes of homologous subunits, α and β. Each subunit combination has distinct pharmacological properties, with both α and β subunits contributing to ligand sensitivity. The $\alpha 3\beta 2$ subunit combination is sensitive to block by 100nM neuronal bungarotoxin (NBT) (98.0±1.9% block) while $\alpha 2\beta 2$ and $\alpha 3\beta 4$ are insensitive. Previous work has shown that the contribution of $\alpha 3$ to the NBT sensitivity of $\alpha 3\beta 2$ can be localized to three distinct sequence segments (84-121, 121-181, 195-215), and that gln198 of $\alpha 3$ (pro in $\alpha 2$) plays a role in determining NBT sensitivity. We have made a series of mutations within these regions of $\alpha 3$, changing residues from what occurs in $\alpha 3$ to what occurs in $\alpha 2$. Changing thr143 of $\alpha 3$, to lys as in $\alpha 2$ (T143K), results in a loss of NBT sensitivity (7.7±4.6% block by 100nM NBT). Amino acid changes in $\alpha 3$ that had no effect on NBT sensitivity include K87I, Q101A, L109H, K111F, K129S and Y139Q. We have also made a mutant of the $\alpha 4$ subunit which forms receptors (with $\beta 2$) partially sensitive to NBT (15.9±1.4 % block by 1μM NBT) and has a proline at position 198, as does $\alpha 2$. Changing pro198 of $\alpha 4$, to gln as in $\alpha 3$, increases NBT sensitivity (87.4±6.8% block by 1μM NBT). Both NBT and dihydro-β-erythroidine (DHβE) are useful probes of the β subunit contribution to competitive antagonist sensitivity. At ACh concentrations approximating the EC20, $\alpha 3\beta 2$ is 50 fold more sensitive to DHβE (IC50≈400 nM) than $\alpha 3\beta 4$ (IC50≈20 μM). 3 μM DHβE effectively blocks $\alpha 3\beta 2$ (90±4.6% block) but has little effect on $\alpha 3\beta 4$ (13.1±5.1% block). Substituting the first 105 N-terminal amino acid residues of $\beta 4$ into $\beta 2$, results in a subunit which forms receptors with $\beta 4$-like ligand sensitivity, i.e. insensitive to both 100 nM NBT and 3 μM DHβE. Substituting in the first 58 residues of $\beta 4$ resulted in intermediate sensitivity to NBT and DHβE (80±1.7% block by 100 nM NBT; 71±6.6% block by 3 μM DHβE). Thus, at least two distinct sections of the β subunit (1-58 and 58-105) are involved in determining sensitivity to competitive antagonism.

NICOTINIC $\alpha 7$ RECEPTORS: ALZHEIMER'S DISEASE TO ALCOHOL ABUSE. CM de Fiebre, RL Papke and EM Meyer. Dept. of Pharmacology and Therapeutics, University of Florida College of Medicine, Gainesville, FL 32610-0267 USA.

While it has been known for many years that [^{3}H]nicotine and [^{125}I]α-bungarotoxin (BTX) bind to sites in brain, only recently have researchers begun to elucidate some of the potential functions of these sites. The $\alpha 7$ cDNA is thought to give rise to a homo-oligomeric channel corresponding to the BTX binding site in brain; the $\alpha 4$ and $\beta 2$ cDNA's are hypothesized to give rise to a multimeric channel corresponding to the brain [^{3}H]nicotine site. Here, we report that the $\alpha 7$ subtype may be important in the interactions between and the resulting co-dependence on nicotine and ethanol and may also be a useful site for selective pharmacotherapy in Alzheimer's Disease. This subtype is particularly sensitive to the antagonistic effects of ethanol, whereas the $\alpha 2\beta 2$, $\alpha 3\beta 2$ and $\alpha 4\beta 2$ subtypes display marginal sensitivity to ethanol. Novel analogs of the naturally-occurring compound, anabaseine, are also being developed as potential agents for the treatment of Alzheimer's Disease. These compounds enhance memory and act as cytoprotective agents in cell culture systems and in animal models of Alzheimer's Disease. Anabaseine derivatives displace the binding of nicotinic ligands and are highly selective agonists at $\alpha 7$ receptors expressed in *Xenopus* oocytes. While both these compounds also act as non-competitive antagonists at $\alpha 4\beta 2$ and $\alpha 7$ receptor subtypes, it is hypothesized that selective $\alpha 7$ activation is responsible for the memory enhancing and cytoprotective actions of these agents. Neuronal $\alpha 7$ receptors may also be a site of action for ethanol in producing dependency on ethanol and co-dependency on ethanol and nicotine.

Rat cDNAs were provided by Dr. Jim Boulter of the Salk Institute and anabaseine compounds were provided by Drs. William Kem and John Zoltewicz of the Univ. of Florida. This work was supported by the Taiho Pharmaceutical Co., & NIA P01 AG10485 (Project 2) to EMM . CmdeF was supported by training grants AG-00196 and AA-07561.

^{125}I-α-BUNGAROTOXIN BINDING PARAMETERS DISCRIMINATE $\alpha 7$ NACHR AGONISTS FROM ANTAGONISTS AND LOBELINE. J. Gordon, S. McCreedy, A. Machulskis and J. Blosser. Fisons Corporation, Rochester, NY 14623.

Nicotinic ligands were profiled in an ^{125}I-α-Btx binding assay to compare the nature of their interaction with the $\alpha 7$ nAChR subtype. Rat hippocampal membranes were coincubated with ^{125}I-α-Btx and test ligands for 2 hrs at 21 ° in 0.1% BSA/Krebs and then collected on GF/C filters pretreated with 1%BSA/0.01%PEI (blanks ≈0.07%). In the presence of 2 mM CaCl$_2$, putative agonists exhibited 5-13 x greater affinities (K$_i$ rank: α-anatoxin < (+/-)-anabasine ≤ (+/-)-anabaseine ≤ dimethylphenylpiperazinium ≤ (-)nicotine ≤ (-)cytisine < (+)nicotine < arecoline ≤ suberyldicholine) in its absence (+0.5 mM EGTA). In contrast, calcium had no effect on antagonist affinities (K$_i$ rank: methyllycaconitine < α-BTX < < d-tubocurarine < dihydro-β-erythroidine). Similar effects of calcium were found by Lukas et al. ((1979) Bioch:18,2384) in rat brain. They also reported Hill coefficients of unity for agonists and antagonists. We found slopes of 1 for antagonists but, in contrast, steep slopes (≈2) for agonists. While the explanation for the difference is unknown, agonists appear to cooperatively bind to the 2 sites/nAChR under our conditions. Slopes were unaffected by calcium. (-)Lobeline, a "mixed" nicotinic ligand *in vivo*, showed mixed properties at the $\alpha 7$ nAChR: a slope of 1 in the absence of calcium but a slope of 2 and a 7x greater affinity in its presence. Agonist and antagonist inhibition of high affinity [^{3}H]nicotine binding to the $\alpha 4\beta 2$ nAChR of rat brain membranes was insensitive to calcium and exhibited slopes of 1. The high affinity states of nAChRs appear to be desensitized states because the affinities of agonists are much greater than their potencies. These results suggest that calcium participates in the desensitization of the $\alpha 7$ but not the $\alpha 4\beta 2$ nAChR subtype.

MUTATIONAL ANALYSIS OF NOVEL RESIDUES IDENTIFIED WITHIN THE BINDING SITE OF *d*-TUBOCURARINE OF *TORPEDO* ACETYLCHOLINE RECEPTOR. Y. Xie, D.C. Chiara and J.B. Cohen* Department of Neurobiology, Harvard Medical School, Boston, MA 02115

The competitive antagonist *d*-tubocurarine (dTC) binds to the two agonist binding sites of the nicotinic acetylcholine receptor (nAChR) from *Torpedo* electric organ and mouse skeletal muscle with significantly different affinities. The high-affinity and low-affinity dTC binding sites are located at the α-γ and α-δ subunit interfaces, respectively. Our photolabeling studies using [^{3}H]-dTC identified previously two non-α residues: γW55 and δW57. Here we report the identification of two additional residues photolabelled by [^{3}H]-dTC: γY111 and γY117. The tyrosine in the γ-subunit of mouse muscle nAChR corresponding to *Torpedo* γY117 has been shown to be important for the selective binding of dimethyl-TC (S. Sine, *PNAS* 90: 9346-9400, 1993). In the aligned *Torpedo* nAChR sequences, the residue which corresponds to γY111 in the δ-subunit is δR113. This difference may also contribute to the difference in dTC binding affinities between the two sites. To examine the role of these two residues in dTC binding, we made γY111R and δR113Y mutant receptors. When expressed in *Xenopus* oocytes, both mutant receptors show similar densities of ^{125}I-α-bungarotoxin (αBgt) binding and amplitudes of ACh-induced currents as wild type receptors. Competition studies of dTC against the initial rate of ^{125}I-αBgt binding to both whole oocytes and oocyte membranes reveal that γY111R mutation causes $\approx$ 3-fold decrease of dTC affinity for the high-affinity binding site, while the δR113Y mutation increases dTC affinity for the low-affinity binding site 2-3 fold. Functional consequences of these mutants in terms of changes of agonist/antagonist sensitivities will be assessed by using the two-electrode voltage-clamp technique.

DIFFERENTIAL BINDING OF NICOTINE AND α-BUNGAROTOXIN TO RESIDUES 173-204 OF THE NICOTINIC ACETYLCHOLINE RECEPTOR α1 SUBUNIT. T. L. Lentz, Dept. of Cell Biology, Yale University School of Medicine, New Haven, Connecticut 06510, USA.

The binding of the agonist L-[^{3}H]nicotine and the competitive antagonist ^{125}I-α-bungarotoxin to overlapping synthetic peptides comprising residues 173-227 of the *Torpedo* nicotinic acetylcholine receptor α1 subunit were compared using a solid phase assay. Both nicotine and α-bungarotoxin bind effectively to peptide 173-204. Equilibrium saturation binding of [^{3}H]nicotine to this peptide revealed a minor binding component with an apparent K_D of 1.9 nM and a major component with a K_D of 1.6 μM. Nicotine bound to α subunit peptides 181-198 and 194-204, less well to 179-192 and 186-196, and did not bind to 173-180 and 205-227. α-Bungarotoxin bound to peptide 186-196, less well to 179-192 and 181-198, and did not bind to 173-180, 194-204, and 205-227. Agonists (nicotine, suberyldicholine, carbamylcholine, and cytisine) effectively competed [^{3}H]nicotine binding to the 173-204 peptide but competed ^{125}I-α-bungarotoxin binding at millimolar concentration and with loss of rank order of potency. The competitive antagonists α-bungarotoxin and α-cobratoxin effectively blocked ^{125}I-α-bungarotoxin binding but only partially competed [^{3}H]nicotine binding. *d*-Tubocurarine competed both α-bungarotoxin and nicotine binding, the latter less effectively. These results indicate that nicotine and α-bungarotoxin preferentially bind to different determinants within residues 173-204. Alternatively, nicotine and α-bungarotoxin could bind to different conformations of the peptide. Both agents appear to interact with common residues, most likely Tyr 190 and Cys 192, in the region of Cys 192 so that there is overlap of binding sites. However, a determinant to the N terminal side of this region, most likely Tyr 189, plays a greater role in α-bungarotoxin binding and a determinant to the C terminal side, most likely Tyr 198, plays a more important role in nicotine binding. *d*-Tubocurarine appears to share determinants of the agonist and antagonist binding sites. Supported by NIH grant NS 21896.

Section 3: Novel nicotinic receptor ligands

NICOTINE: STRUCTURE-AFFINITY STUDIES; DEVELOPMENT OF NOVEL AGENTS.
W. Fiedler, M. Dukat, M.I. Damaj, B.R. Martin and R.A. Glennon. Departments of Medicinal
Chemistry and Pharmacology, Medical College of Virginia, Virginia Commonwealth University;
Richmond, VA 23298.

Few agents (agonists) bind at [³H]nicotine-labeled nicotine receptors with an affinity comparable
to that of nicotine. We initially approached this problem by undertaking a systematic structure-
affinity study to determine the contribution to binding of various aspects of the nicotine molecule
and of nicotine-related derivatives. For example a series of aryl-substituted and unsubstituted,
primary, secondary, and tertiary amine derivatives of 3-(aminomethyl)pyridine (1) were prepared
and examined. The finding that 1 (R = Me, R'= Et, X = H) binds with good affinity (Ki = 28

nM) led to the synthesis of several conformationally-restricted analogs, such as 2 and 3 (Ki = 85 and
12 nM, respectively). With the availability of such structure-affinity and conformational data, it
should now be possible to optimize affintiy by design and synthesis of new compounds.
(Supported in part by funding from TDC/CIT.)

**N-SUBSTITUTED NICOTINE ANALOGS, A NEW CLASS OF NICOTINIC RECEPTOR
ANTAGONIST.** L.H. Teng, A. Ravard, S.T. Buxton, P.A. Crooks and L.P. Dwoskin. College of
Pharmacy, University of Kentucky, Lexington, KY, 40536 USA.

A series (1a-1f and 2) of quaternary ammonium derivatives of S(-)nicotine (NIC) were synthesized
and evaluated for nicotinic antagonist activity. Rat striatal slices were preloaded with [³H]dopamine
(DA) and superfused with Krebs' buffer containing 10 μM nomifensine and 10 μM pargyline. After
60 min of superfusion, slices were exposed to test compounds (1-100 μM) included in the superfusion
buffer to determine their ability to evoke [³H]DA release. Subsequently, NIC (10 μM) was added to
the buffer, and the ability of the test compound to inhibit NIC-evoked [³H]DA release determined.
Compounds 1a and 1c did not exhibit nicotinic antagonist activity; and 1c (100 μM) markedly increased
[³H]DA release. Compounds 1b, 1f and 2 exhibited weak antagonist activity (low potency and efficacy).
At 100 μM, compounds 1d and 1e completely blocked the NIC-evoked [³H]DA release. IC50s for 1d
and 1e were 30 and 10 μM, respectively. Both 1d and 1e were devoid of [³H]DA releasing properties
even at 100 μM. Also, 1e did not alter [³H]DA release evoked by electrical field-stimulation. From
structure-activity considerations, NIC derivatives containing a 3-carbon substituent on the pyridyl
nitrogen, which affords a water soluble quaternary ammonium salt, exhibit antagonistic activity at
nicotinic receptors. This is the first report of a new class of neuronal nicotinic receptor antagonists.
Future work will determine the selectivity of these antagonists at specific nicotinic receptor subunits.
(Supported by a grant from the Tobacco and Health Research Institute, Lexington, KY.)

	R₁	R₂	X
a	Me	Me	O
b	Me	Me	H₂
c	Me	H	H₂
d	Propyl	Me	H₂
e	Allyl	Me	H₂
f	Allyl	Allyl	H₂ (racemic)

**TRICYCLOPINATE HCL–A NEW SYNTHETIC COMPOUND WITH BOTH
MUSCARINIC AND NICOTINIC ANTAGONISTIC ACTIVITIES.** Chuan–Gui Liu, Hai
Wang, De–Lu Zhao, Zhan–Guo Gao, Wen–Yu Cui, Shu–Ping Zhang, Qing–Suo Qiao and
Yun–Zhang Ran. Institute of Pharmacology and Toxicology, P.O. Box 130, Beijing 100850,
China

In recent years a number of central cholinergic antagonists were synthesized in our Institute.
Among them Tricyclopinate HCl (TCPN) was found to show both antimuscarinic and
antinicotinic activities. The ability of TCPN to displace (³H) QNB binding to membranes of rat
cerebral cortex, heart, submaxillary gland and guinea pig ileum longitudinal muscle (IC₅₀:2.24,
2.72, 1.78 and 1.59 nM respectively) was just at the same level as QNB (IC₅₀:2.65, 1.27, 1.58 and
3.00 respectively). These results were confirmed by using (³H) TCPN as radioligand. On the other
hand TCPN could prevent iv nicotine induced convulsion in mice with ED₅₀ of 0.59 mg / kg iv
and could shift the dose–response curve of nicotine for producing mice convulsion to the right in
parallel manner. Furthermore TCPN could depress the nicotine induced contraction of
longitudinal muscle of guinea pig ileum with ED₅₀ of 64.2 nM in vitro. The ability of TCPN to
displace (³H) Nicotine binding to membranes of brain of rat is now under investigating. It is expected that through QSAR analysis a number of compounds with high central antinicotinic and
low / no central antimuscarinic activities may be synthesized.

P 21

α-CONOTOXIN ImI, A SELECTIVE LIGAND FOR NEURONAL nAChRs. J. M. McIntosh[1,2], D. S. Johnson[3], D. Yoshikami[2], E. Mahe[3], D. B. Nielsen[2], J. E. Rivier[3], W. R. Gray[2] and B. M. Olivera[2]. Depts. of Psychiatry[1] and Biology[2], University of Utah, Salt Lake City, UT; Salk Institute[3], La Jolla, CA.

A small peptide has been isolated which targets to nicotinic acetylcholine receptors (nAChRs). The peptide potently blocks the neuromuscular receptor in frogs but not mice. Like α-Bungarotoxin it produces complex seizures when injected centrally into rats, suggesting that it targets to neuronal nAChRs in mammals. *Xenopus* oocyte expression studies indicate the peptide does selectively target a specific central nAChR. The peptide is a highly divergent α-conotoxin isolated from the worm-hunter, *Conus imperialis* and has thus been named α-conotoxin ImI (α-CTx-ImI). Large snake α-toxins (~80 amino acids) target both mammalian muscle and central nAChRs. In contrast, α-CTx-ImI is only 12 residues yet is more selective. Its small size has allowed for total chemical synthesis; native and synthetic toxin are structurally and functionally indistinguishable. Central nAChRs appear important in a number of neuropsychiatric disorders. α-CTx-ImI represents a unique tool for the study these receptors. Cone venoms in general may prove to be a rich source of selective neuronal nAChR-targeted peptides.

P 22

THE MARINE TOXIN ANABASEINE IS A POTENT NICOTINIC AGONIST. Kem, W.R., Mahnir, V.M., Lin, B., Lingle*, C.J., Papke, R.L., and Prokai-Tatrai, K. Dept. of Pharmacology and Therapeutics, University Florida College of Medicine, Gainesville, FL 32610; *Anesthesiology Research Unit, Dept. of Anesthesiology, Washington University, St. Louis, MO 63110.

Stimulation of brain nicotinic receptors may be a means of ameliorating some of the cognitive deficits associated with loss of central nicotinic receptors occurring in some neurodegenerative diseases (Alzheimers and Parkinsons). Because nicotine stimulates all receptor subtypes, toxic side effects may limit its use for this purpose. Anabaseine, (2-(3-pyridyl)-3,4,5,6-tetrahydropyridine), is a constituent of several nemertine worm venoms (Kem, 1971, 1988). We have investigated the ability of this compound, relative to nicotine, to activate various neuronal and muscle receptor subtypes. Anabaseine administered icv was 3X less potent that nicotine in causing prostration. The anabaseine IC_{50} for inhibition of rat brain high nicotine affinity binding (measured with [3H]MCC) was 10X higher than for nicotine, whereas the IC_{50} for inhibition of alpha-bungarotoxin binding was about 3X lower than for nicotine. On alpha4-beta2 nicotinic receptors expressed in the *Xenopus* oocyte anabaseine acted as an apparent partial agonist which may be due to mixed agonist and noncompetitive inhibitor effects. The PC12 maximum rubidium permeability increase by anabaseine was very similar to that of nicotine, but the EC_{50} was 3X higher than for nicotine. On the rat colon longitudinal muscle-myenteric plexus preparation, anabaseine also acted as an apparent partial agonist. The neuromuscular potency of anabaseine was about 5X higher than for nicotine. Single channel measurements on cultured BC3H1 myotubules revealed two actions of anabaseine: activation of the nicotinic receptor and channel blockade. Relative to the chemically related tobacco alkaloid anabasine, anabaseine is a much more potent nicotinic agonist on neuromuscular nicotinic receptors. (Partially supported by Taiho Pharmaceutical Co., Ltd.)

CHOLINERGIC CHANNEL ACTIVATORS (ChCAs) FOR THE POTENTIAL
TREATMENT OF CNS DISORDERS
S. P. Americ, J.P. Sullivan, and M. Williams. Neuroscience Research (D-47W) Pharmaceutical
Products Division, Abbott Laboratories, Abbott Park, IL, 60064-3500
     With the plethora of receptor targets evolving as the result of nicotinic acetylcholine receptor
(nAChR) cloning it is possible that new molecular entities selective for subtypes of nAChRs can be
developed which are potentially free of the side effect liabilities associated with (-)-nicotine. Recent data
also suggests that neuronal nAChR function can be enhanced at sites distinct from where (-)-nicotine
interacts on the α subunit. nAChR agonists whose actions may occur via the selective interaction with
central nAChRs subtypes, and allosteric modulators that indirectly affect ligand-gated nAChR function
encompass a broader class of compounds which can be termed cholinergic channel activators (ChCAs).
Functionally, ChCAs may enhance central neuronal nAChR mediated transmission while substantially
reducing the side-effect liabilities normally associated with (-)-nicotine. ChCAs lacking cardiovascular or
other CNS side effects associated with (-)-nicotine may thus represent a potential novel therapeutic
approach to ameliorate many of the CNS deficits accompanying AD or other related disorders. ABT 418
[(S)-3-methyl-5-(1-methyl-2-pyrrolidinyl) isoxazole hydrochloride] is a selective prototypic ChCA
currently in development for the treatment of AD (see accompanying abstracts). ABT-418 may be a safe
and effective ChCA for the potential treatment of the cognitive and emotional impairments associated with
AD. The therapeutic potential of ChCAs for neuroprotection, smoking cessation, anxiety disorders,
schizophrenia, attentional deficit disorder, Tourette's syndrome, analgesia and depression will be
reviewed.

INTERACTION OF ABT-418, NICOTINE, AND THEIR NOR- AND EPI- ANALOGS WITH
CHOLINERGIC CHANNEL RECEPTORS: BINDING, FUNCTIONAL ACTIVITY, AND
MOLECULAR MODELING STUDIES. M. W. Holladay, J. T. Wasicak, D. Donnelly-Roberts, D. J.
Anderson, P. Pavlik, Y. C. Martin, D. S. Garvey, J. P. Sullivan, and S. P. Americ Neuroscience Research,
Abbott Laboratories, Abbott Park, IL 60064-3500

     ABT-418 (3-methyl-5-(1-methyl-2(S)-pyrrolidinyl)-isoxazole) is a novel cholinergic channel activator
originally designed as a bioisosteric analog of (S)-nicotine,[1] but which has been shown to possess distinct
properties.[1,2] Since (R)-nicotine and both isomers of nornicotine also are active at cholinergic channels,
the enantiomer of ABT-418 and both of the corresponding N-desmethyl analogs have been studied in
comparison with the corresponding nicotine analogs. Radioligand binding and ligand-gated ^{86}Rb$^+$ efflux
in several preparations will be reported, together with the results of molecular modeling studies in which
the possible bioactive conformations of the ABT-418, nicotine, and enantiomeric and N-desmethyl
analogs are examined. A new method for optical resolution of nornicotine also will be described.

1. Garvey, D.S., Wasicak, J. T., Decker, M. W., Brioni, J. D., Buckley, M. J., Sullivan, J. P., Carrera, G.
M., Holladay, M. W., Americ, S. P., Williams, M. "Novel Isoxazoles Which Interact with Brain
Cholinergic Channel Receptors Have Intrinsic Cognitive Enhancing and Anxiolytic Activities," *J. Med.
Chem.* **1994**, *in press.*

2. Americ, S. P., Sullivan, J. P., Briggs, C. A., Donnelly-Roberts, D, Anderson, D, Raskiewicz, J.,
Hughes, M., Cadman, E., Adams, P., Garvey, D., Wasicak, J., Williams, M. "ABT 418: A Novel
Cholinergic Ligand With Cognition Enhancing and Anxiolytic Activities: I. In Vitro Characterization," *J.
Pharmacol. Exp. Therap.* **1994**, *in press.*.

ABT-418: IN VITRO PROPERTIES OF A NOVEL CHOLINERGIC CHANNEL
ACTIVATOR (ChCA) FOR THE POTENTIAL TREATMENT OF ALZHEIMER'S
DISEASE. J. P. Sullivan, D.J. Anderson, D.Donnelly-Roberts, G. Wilkie[1], Susan Wonnacott[1],
D. S. Garvey, M.Williams, and S.P. Americ. Neuroscience Research (D47W), Abbott Laboratories,
Abbott Park, IL, 60064 and [1]Dept. Biochemistry, University of Bath, Bath, England
     The potential therapeutic usefulness of (-)-nicotine, the prototype agonist for neuronal nicotinic
acetylcholine receptors (nAChRs) is severely hampered by it's dose-limiting side-effects. Compounds
that selectively activate nAChR subtypes to normalize CNS functions but lack the side-effect liabilities of
(-)-nicotine may, therefore, lead to more effective therapeutic agents (Americ et al., see accompanying
abstract). In the present study the *in vitro* properties of ABT-418 [(S)-3-methyl-5-(1-methyl-2-
pyrrolidinyl) isoxazole), a novel cholinergic channel activator, are described. ABT-418 was a potent
inhibitor of [^{3}H](-)-cytisine binding to nAChRs in rat brain (K_i = 3.0 ± 0.4 nM) but was significantly
less potent (K_i > 10,000 nM) in 37 other receptor/neurotransmitter uptake/enzyme binding assays
including those for the alpha-bungarotoxin sensitive nAChR subtypes ($\alpha\beta\delta\gamma$ and $\alpha7$). In IMR 32 cells,
cation efflux studies indicated that ABT-418 was a less potent activator of ganglionic-like nAChRs than
(-)-nicotine (EC$_{50}$: ABT-418, 100 ± 20 µM; (-)-nicotine, 21 ± 5 µM); an effect prevented by the nAChR
channel blocker, mecamylamine. ABT-418 was also approximately ten-fold less potent than (-)-nicotine
in evoking [^{3}H]dopamine release from rat striatal slices (EC$_{50}$: ABT-418, 380 ± 20 nM; (-)-nicotine, 40
± 6 nM), but was as efficacious and only slightly less potent than (-)-nicotine in stimulating [^{3}H]ACh
release from rat hippocampal synaptosomes. In contrast, ABT-418 was more efficacious than (-)-
nicotine in stimulating cation efflux from mouse thalamic synaptosomes, thought to reflect activation of
the $\alpha4\beta2$ nAChR subtype- the major subtype in rodent brain. The ability of ABT-418 to differentially
activate nAChR subtypes may account for the substantial separation between the cognitive
enhancement/anxiolytic benefits and the reduced CNS side-effect liabilities observed in vivo (Decker et
al., see accompanying abstract).

EFFECTS OF ABT-418 ON NICOTINIC RECEPTOR MEDIATED ^{86}Rb$^+$ EFFLUX FROM MOUSE BRAIN SYNAPTOSOMES. M.J.Marks, S.F.Robinson, and D. Donnelly-Roberts*. Univeristy of Colorado, Boulder, CO and *Abbott Laboratories, Abbott Park, IL.

ABT-418 [(S)-3-methyl-5(1-methyl-2-pyrrolidinyl)-isoxazole hydrochloride], a potent nicotinic agonist that exhibits fewer side effects than nicotine (Arneric et al., Decker et al. J. Pharmacol. Exp. Ther. 1994, in press), was evaluated for its ability to stimulate the efflux of ^{86}Rb$^+$ from synaptosomes prepared from the thalamus of C57BL/6 mice, in order to determine the properties of this compound at receptors apparently corresponding to high affinity [^{3}H]nicotine binding sites (possibly the $\alpha4/\beta2$ receptor subtype) (Marks et al., J.Pharmacol.Exp. Ther. 264:542, 1993). Synaptosomes loaded with ^{86}Rb$^+$ were superfused with buffer containing 5 mM Cs$^+$, 100 nM tetrodotoxin, and 0.1% bovine serum albumin. ABT-418 stimulation of ^{86}Rb$^+$ efflux was concentration-dependent and saturable with an EC$_{50}$ of 6 μM and a maximum response 140% that of 10 μM nicotine. The response measured for 1 min using 30 μM ABT-418 was inhibited more than 95% by 10 μM mecamylamine. Prolonged exposure to stimulating concentrations of ABT-418 (1 μM - 100 μM) desensitized ^{86}Rb$^+$ efflux. The EC$_{50}$ value for desensitization was comparable to that for stimulation of efflux. Exposure of the synaptosomes to non-stimulating concentrations of ABT-418 also desensitized the efflux stimulated by 10 μM nicotine (IC$_{50}$ = 300 nM). The desensitization rate for 300 nM ABT-418 (0.27 min^{-1}) was less than the maximum rate observed for activating concentrations (1.8 min^{-1}). Synaptosomes that had been desensitized by exposure to either nonstimulating (300 nM) or stimulating (30 μM) concentrations of ABT-418 partially recovered responsiveness after the termination of treatment. The recovery from desensitization after 300 nM ABT-418 was faster than that after 30 μM (k = 0.34 min^{-1} and 0.19 min^{-1}, respectively). The kinetics of desensitization observed after exposure to either stimulating or nonstimulating concentrations of ABT-418 were comparable to those measured previously for nicotine. The results of these studies indicate that ABT-418 is a potent agonist at nicotinic receptors (possibly the $\alpha4/\beta2$ subtype) in the CNS. While ABT-418 had a lower affinity (EC$_{50}$ = 6 μM) than nicotine (EC$_{50}$ = 0.8 μM) for these receptors, it elicited a greater response than nicotine.

[^{3}H]ABT-418: RECEPTOR BINDING PROPERTIES OF A NOVEL CHOLINERGIC CHANNEL LIGAND. James P. Sullivan, David J. Anderson, James R. Pauly[1],Michael Williams, and Stephen P. Arneric. Neuroscience Research (D47W), Abbott Laboratories, Abbott Park, IL, 60064 and [1] Dept. Pharmacology, Medical College Georgia, Augusta, GA, 30912.

ABT-418 is a novel cholinergic channel ligand with cognitive enhancing and anxiolytic activities (see Decker et al, accompanying abstract). In the present studies, the binding characteristics of [^{3}H] ABT-418 were investigated. [^{3}H]ABT-418 was found to bind with high affinity (K$_D$ = 3.1 ± 0.1 nM; n=5) to membranes prepared from whole rat brain. Binding of [^{3}H] ABT-418 was characterized by rapid association (t$_{1/2}$ = 1.4 ± 0.3 min.) and dissociation (t$_{1/2}$ = 2.9 ± 0.4 min.) half-times. Maximal binding of [^{3}H]ABT-418 was directly compared to the neuronal nicotinic acetylcholine receptor (nAChR) radioligand, [^{3}H](-)-cytisine, in several brain regions (cortex, hippocampus, thalamus, striatum and cerebellum). Although the regional distributions were similar, the number of receptors labeled by [^{3}H]ABT-418 was significantly lower (18-33% ; p < 0.05) in the cortex, striatum and cerebellum suggesting that ABT-418 may be binding to a subpopulation of nAChRs labeled by [^{3}H](-)-cytisine. Nonetheless, the pharmacology of the binding site in whole rat brain is similar to that found using [^{3}H](-)-cytisine. The nAChR agonists, (-)-nicotine, (-)-cytisine and (±)-epibatidine displayed a high affinity (K$_i$ = 0.8 ± 0.1 nM, 0.2 ± 0.1 nM and 0.05 ± 0.01 nM, respectively) for [^{3}H]ABT-418 binding sites while among nAChR antagonists examined, only dihydro-β-erythroidine competed with high affinity (K$_i$ = 32 ± 1.5 nM). Thus, [^{3}H]ABT-418 binds with high affinity to sites in brain that have the pharmacological characteristics of neuronal, but not α-bungarotoxin sensitive nAChRs. Furthermore the results suggest that this agent may be binding to a subpopulation of the neuronal receptors labeled by [^{3}H](-)-cytisine in some brain regions which may in part explain the *in vivo* differences in pharmacological profile between ABT-418 and classical nicotinic agonists.

ABT-418: IN VIVO PROFILE OF A NOVEL CHOLINERGIC CHANNEL ACTIVATOR (ChCA) FOR THE POTENTIAL TREATMENT OF ALZHEIMER'S DISEASE (AD) M.W. Decker, J.D. Brioni, M.J. Buckley, P. Curzon, A.B. O'Neill, D.J.B. Kim, M. Majchrzak, K. Marsh, S. Quigley, A.D. Rodrigues, R. Radek, J.P. Sullivan, and S.P. Arneric. Neuroscience Research (D-47W) Pharmaceutical Products Division, Abbott Laboratories, Abbott Park, IL, 60064-3500

ABT-418 has been evaluated in a series of animal test paradigms to assess cognition enhancement. ABT-418 had a positive effect at 0.062 μmol/kg on the retention of *inhibitory avoidance* with pre-training injections in mice and at 0.62 μmol/kg with post-training injection in rats, i.p. (-)-Nicotine produced similar effects at 3 to 10-fold higher doses. The cognition enhancing activity of ABT-418 in this model in mice was prevented by the centrally acting nAChR channel blocker, mecamylamine (5 μmol/kg, i.p.). The effects were stereoselective since the (R)-enantiomer of ABT-418, A-81754, was without effect. Enhancement of performance was maintained in aged (20 months) rats following continuous infusion with osmotic minipumps over an 11-day treatment period. The *Morris water maze* paradigm was used to measure spatial memory in the medial septal lesioned rat in which cholinergic innervation of hippocampus is reduced. ABT-418, given i.p. (0.19 and 1.9 μmol/kg) restored performance in a dose-related manner essentially to control levels. ABT-418 demonstrated anxiolytic-like activity in both mice and rats in the *elevated plus maze* model of anxiety at doses of 0.19 and 0.62 μmol/kg, i.p., respectively--an effect blocked by mecamylamine (15 μmol/kg). ABT-418 was approximately 15-fold more potent than diazepam in mice, but was less efficacious in eliciting anxiolytic-like activity. Nonetheless, in contrast to diazepam, ABT-418 did not potentiate ethanol-induced narcosis, nor did it impair rotorod performance in the effective dose range. ABT-418 (0.62 μmol/kg, i.p.) also reduced the anxiety elicited by withdrawal from 14 days of (-)-nicotine treatment by minipump infusion. While ABT-418 had approximately the same potency as (-)-nicotine in memory tasks, the compound was less potent than (-)- nicotine in producing hypothermia, seizures and death in rodents. In contrast to (-)- nicotine, ABT-418 had significantly less emetic and pressor liability in dog. In rodent, dog and monkey ABT-418 demonstrated substantial transdermal bioavailability, yet poor oral bioavailability due to rapid metabolism.

IMPROVEMENT IN PERFORMANCE OF A DELAYED MATCHING-TO-SAMPLE TASK BY MONKEYS GIVEN ABT-418, A NOVEL nAChR ACTIVATOR FOR MEMORY ENHANCEMENT J.J. Buccafusco, W. J. Jackson, A.V. Terry, Jr., K.C. Marsh, M.W. Decker and S. P. Americ. Dept. Pharmacol. Toxicol. (JJB, AVT), and Department of Physiol. Endocrinology (WJJ), Medical College of Georgia, and Dept. V. A. Med. Ctr., Augusta, Georgia (JJB) and Neuroscience Discovery, Abbott Laboratories, Abbott Park, Illinois (KCM, MWD and SPA).

ABT-418, a newly characterized centrally-acting nicotinic acetylcholine receptor (nAChR) activator, was evaluated for its ability to improve performance in a delayed matching-to-sample (DMTS) task by mature macaques well trained in the task. Previous studies in rodent models have indicated that ABT-418 shares the memory/cognitive enhancing actions of nicotine, but without many of nicotine's adverse side effects. The DMTS task provides a measure both of general cognitive function and of recent memory. It was hypothesized that doses of ABT-418 would enhance the monkeys' ability to correctly perform the DMTS task. Intramuscular administration of ABT-418 significantly enhanced DMTS performance at low (nmol/kg) doses. In fact, the drug was slightly more potent than nicotine in this regard, and all 8 animals tested in this study exhibited enhanced performance at one or more doses. Unlike nicotine, however, the enhanced performance to ABT-418 was observed only on the day of administration, whereas, nicotine's effects were still evident on the next day's testing session. This difference in the apparent duration of action between the two drugs could not be attributed to a longer plasma half-life for nicotine. ABT-418 produced a delay-dependent increase in DMTS performance with the greatest improvement observed for the longest delay interval. In animals repeatedly tested with their individualized "best dose", DMTS performance increased on average by 10.1 ± 3.5 percentage points correct, which was equivalent to an increase of 16.2% over baseline performance. ABT-418 did not significantly affect response times, i.e., latencies to make a choice between stimuli, or latencies to initiate new trials. Finally, single daily administration of the individualized best dose in 3 monkeys over a period of 8 days did not result in a fully sustained enhancement of DMTS performance, but neither was the data consistent with the development of significant tolerance to the drug's mnemonic actions. No overt toxicity or side effects to acute or chronic administration of the drug were noted. Thus, ABT-418 represents a new class of nAChR activator designed for the treatment of human dementias having a low profile of potential toxicity.

AUTORADIOGRAPHIC COMPARISON OF [3H]-CYTISINE AND [3H] ABT-418 BINDING IN RAT BRAIN. J.R. Pauly*, S.P Americ, M. Williams and J.P. Sullivan. *Dept of Pharmacology and Toxicology, Medical College of Georgia, Augusta, GA 30912 and Abbott Laboratories, Neuroscience Research (D-47W), Abbott Park IL 60064.

With the possible exception of alpha bungarotoxin, ligands that identify neuronal nicotinic cholinergic receptors based on their alpha and beta subunit composition have not been identified. In homogenate receptor binding assays, the novel cholinergic channel activator ABT-418 demonstrates a significantly lower Bmax as compared to [3H]-cytisine binding in several brain regions, suggesting that this ligand may bind to a subpopulation of nicotinic receptors (Sullivan, et al., 1994). The purpose of the present study was to compare the binding of [3H]cytisine and [3H]ABT-418 in sections of rat brain using autoradiography. Adjacent cryostat sections were pre-incubated in KRH buffer for 30 min (4°C) and then transferred to fresh buffer containing saturating levels of [3H]cytisine or [3H]ABT-418 and incubated for 2 hr (4°C). The samples were then washed twice in KRH buffer, diluted KRH buffer and deionized water (10 sec per wash: 4°C). The distribution of [3H]ABT-418 binding was similar to [3H]cytisine binding in many brain regions including most thalamic nuclei (e.g. anteroventral, reunions, rhomboid), the bed nucleus of the stria terminalis, medial habenula, subiculum, dorsolateral geniculate, and the interpeduncular nucleus. However, several brain regions labeled prominently by [3H]cytisine had significantly lower binding of [3H]ABT-418. These brain regions included the caudate putamen, olfactory tubercle, superior colliculus, substantia nigra, cerebellum and certain cortical regions. These data suggest that ABT-418 may bind to a subset of the receptors identified by [3H]-cytisine binding.

Section 5: Epibatidine, a novel nicotinic receptor agonist

P 31

EPIBATIDINE: A HIGH - AFFINITY NICOTINE RECEPTOR LIGAND.
M. Dukat, D. Dumas, M.I. Damaj, W. Glassco, E.L. May, B.R. Martin and R.A. Glennon.
Departments of Medicinal Chemistry and Pharmacology, Medical College of Virginia, Virginia
Commonwealth University; Richmond, VA 23298.

It is speculated that nicotine receptor ligands might be of therapeutic benefit for the treatment of
obesity, anxiety, and memory loss. With the exception of nicotine, few selective high-affinity agents
exist. (-)Epibatidine (1), isolated from Ecuadoran frogs, shows structural resemblence to (-)nicotine
(2). Molecular modeling studies reveal that the N to N distance in 1 (5.5 Å) is greater than that
found in 2 (4.9 Å). This distance exceeds that previously considered optimal for the nicotine
receptor pharmacophore (4.8 ± 0.3 Å) by about half a bond length. Nevertheless 1 binds (Ki = 0.055
nM) at [^{3}H]nicotine-labeled receptors with higher affinity than 2 (Ki = 1.5 nM). Although our
studies are still in progress, molecular superimposition and preliminary results with various structurally modified analogs suggest that it is the conformationally constrained nature of 1 that accounts
for its high affinity. For example, 6-chloronicotine binds only with slightly higher affinity (Ki = 0.6
nM) than nicotine suggesting, at least for nicotine, that the chloro substituent is tolerated but does
not contribute significantly to binding. Excision of the ethano bridge of deschloroepibatidine, to
yield isonornicotine (Ki = 12.5 nM), results in reduced affinity. Isonicotine binds with similar (Ki
= 7.3 nM) affinity. The bridged azabicycloheptane system likely fixes the structure of epibatidine
in a highly receptor-preferred conformation. The pharmacophore for nicotine receptor binding will
require reconsideration and should take into account the longer N-N distance of epibatidine. The
N-N distance in epibatidine may be optimal, whereas that in isonornicotine is longer, and that for
nicotine shorter. (Supported in part by funding from TDC/CIT.)

P 32

EPIBATIDINE AND RELATED ANALOGS COMPETE WITH ^{3}H-CYTISINE
WITH HIGH AFFINITY FOR BINDING TO RAT BRAIN CORTICAL
MEMBRANE PREPARATIONS. D.M. Wypij[1] and T.Y. Shen[2]. [1]CytoMed,
Inc., 840 Memorial Drive, Cambridge, MA 02139, USA and [2]Department of
Chemistry, University of Virginia, Charlottesville, VA 22901, USA

Epibatidine, an alkaloid isolated from skin extracts of an Ecuadorian poison
frog, is a potent non-opioid analgesic. We have previously shown that the
analgesic effect of epibatidine can be blocked by the nicotinic antagonist
mecamylamine. To extend our initial pharmacological observations, these
compounds were examined for their ability to bind to neuronal (rat cortex)
nicotinic receptors. IC$_{50}$ values for racemic epibatidine and analogs were
compared to (-)-nicotine (relative value of 1) for displacement of ^{3}H-cytisine
using a one-site model of ligand-receptor interaction. The rank order of
potency was: epibatidine (11) > N-methyl epibatidine >> nicotine (1) >>
"endo" epibatidine (0.062) > N-acetyl epibatidine (0.015). Epibatidine competes
with cytisine for binding to rat cortical membrane preparations with high
affinity (IC$_{50}$ = 94 pM, K$_i$ = 44 pM) in a reversible manner.

P 33

**PHARMACOLOGICAL EFFECTS OF EPIBATIDINE, A POTENT NICOTINIC
AGONIST.**
M. I. Damaj, K. R. Creasy, J. Rosecrans and B. R. Martin. Department of Pharmacology/Toxicology,
Virginia Commonwealth University/Medical College of Virginia, Richmond, VA 23298-0613

Epibatidine, an alkaloid originally isolated from frog skin (J. Am. Chem. Soc. 1992, 114, 3475) and
structurally related to nicotine, has been found to possess a potent analgesic effect in mice using the hot
plate assay. We therefore, investigated the effect of its d and l-enantiomers (Gift from Merck Sharp &
Dohme) in the tail flick test and evaluated them in behavioral assays and receptor binding. Both d- and l-
epibatidine enantiomers competed with high affinity (K$_i$ = 54.7 and 55 pM, respectively) for [^{3}H]-
nicotine binding. Systemic administration (s.c.) of both enantiomers in male ICR mice induced an
antinociceptive effect and decreased locomotor activity and body temperature in a dose-dependent manner
as indicated in the table (ED$_{50}$'s expressed as µg/kg).

Drug	Analgesia	Hypomotility	Hypothermia
l-nicotine	1200	650	1000
l-epibatidine	6.1	1.8	1.9
d-epibatidine	6.6	1.2	3.3

The antinociceptive, hypothermic and locomotor effects of epibatidine enantiomers were completely
blocked by mecamylamine (1 mg/kg), a centrally active nicotinic receptor antagonist, whereas
hexamethonium (1 mg/kg), a peripheral nicotinic receptor antagonist, and naloxone (1 mg/kg) failed to
block their antinociceptive and hypothermic effects. When d and l-epibatidine were evaluated in rat drug
discrimination, they generalize to nicotine in a dose-dependent manner with an ED$_{50}$ of 0.93 and 1.0
µg/kg, respectively, after s.c. administration. This effect was blocked by mecamylamine (1 mg/kg) but
was insensitive to hexamethonium pretreatment. These results show no significant enantioselectivity of
epibatidine optical isomers for nicotinic receptors and provide strong evidence for epibatidine as a very
potent nicotinic receptor agonist. (Supported by PHS grant #DA-05274)

(±)EPIBATIDINE ELICITS A DIVERSITY OF NICOTINIC RECEPTOR-MEDIATED EFFECTS. J.P. Sullivan, A.W. Bannon, D. Donnelly-Roberts, D.J. Anderson, M. Piattoni-Kaplan, M. Gopalakrishnan, M.W. Decker and S.P. Arneric. Neuroscience Research (D-47W), Abbott Laboratories, Abbott Park, IL, 60064.

(±)-Epibatidine, exo-2-(6-chloro-3-pyridyl)-7-azabicyclo-[2.2.1] heptane, is a novel, potent analgesic agent acting through nicotinic acetylcholine receptor (nAChR) mechanisms. In the present study the in vitro and in vivo pharmacological properties of this novel ligand were further characterized. (±)-Epibatidine displaced $[^3H](-)$-cytisine binding to the $\alpha 4\beta 2$ subtype of the nAChR in rat brain membranes with high affinity ($K_i = 43$ pM). The compound was approximately 5000-fold less potent ($K_i = 230$) in displacing $[^{125}I]$ α–BgT binding from the α–BgT-sensitive nAChR subtype present in rat brain but was a potent inhibitor ($K_i = 2.7$ nM) of $[^{125}I]$ α–BgT to the nAChR subtype in *Torpedo* electroplax which is similar to that present in the neuromuscular junction. Functionally, (±)-epibatidine enhanced $^{86}Rb^+$ flux in IMR 32 cells with an EC_{50} value of 7 nM, being some 3000-fold more potent than (-)-nicotine (EC_{50} value = 21,000 nM), and was approximately 150-fold more potent (EC_{50} value = 0.4 nM) than (-)-nicotine (EC_{50} value = 60 nM) in increasing $[^3H]$dopamine release from rat striatal slices. Both functional effects were blocked by the nAChR channel-blocker, mecamylamine (100 μM). *In vivo*, (±)-epibatidine was 300-1000 times more potent than (-)-nicotine in reducing body temperature and locomotor activity in mice. Substantial reductions in both measures were evident at 0.019 μmol/kg, effects that were also blocked by mecamylamine (15 μmol/kg, i.p.). (±)-Epibatidine, at doses ranging from 0.05 μmol/kg to 0.1 μmol/kg, i.p., elicited significant analgesic response on the hot plate, an effect blocked by mecamylamine. Collectively, these results demonstrate that (±)-epibatidine is a potent nAChR agonist in vitro and in vivo with differential activity to evoke responses elicited by putative subtypes of nAChRs. Therefore, (±)-epibatidine may serve as a novel pharmacological tool to further probe nAChR function.

EPIBATIDINE IS A MORE POTENT DESENSITIZER OF NEURONAL NICOTINIC RECEPTORS THAN NICOTINE. R.Loring, T.McHugh, X.Zhang and J.McKay. Dept. Pharmaceutical Sciences, Northeastern University, Boston, MA 02115, USA.

Epibatidine (EB), exo-2(6-chloro-3-pyridyl)-7-azabicyclo [2.2.1] heptane (Spande et al., 1992, *J.Am.Chem.Soc.* **114**:3475), is a natural product with chemical features similar to nicotine, and is reported to be a potent nicotinic analgesic (Qian et al., 1993, *Eur.J.Pharm.* **250**:R14). EB effects were investigated by electrophysiological recordings from perfused chick retina, and by radioligand assays for various nicotinic receptor subtypes. A single 2 sec application of EB (10 μM) caused a curare-sensitive depolarization, but subsequent depolarizations due to applications of the nicotinic agonist dimethylphenylpiperazinium (DMPP, 300 μM, 2 sec) were diminished to less than 30% of control. More than 30 min was required for recovery of DMPP responses after blockade by the single EB application. EB did not affect depolarizations by the glutamatergic agonists kainic acid or N-methyl-d-aspartic acid. 10 μM nicotine is the minimum dose to completely desensitize DMPP responses in retina when applied continuously for 20 min, but DMPP responses recover 50% from nicotine blockade within ≈20 min. In contrast, a hundred-fold lower EB concentration (100 nM, 20 min) completely blocks DMPP responses, and requires at least 1 h for 50% recovery. Preliminary studies suggest that d-tubocurarine (300 μM) prevents the long-term desensitization by EB. EB displaces ^3H-cytisine binding to $\alpha 4\beta 2$ receptors ($K_i = 3$nM) and ^{125}I-α-bungarotoxin binding to $\alpha 7$-containing receptors ($K_b = 22$nM) immunoisolated from chick brain. The corresponding K_i's for nicotine were 20nM for ^3H-cytisine and 140nM for ^{125}I-α-bungarotoxin. In addition, EB blocked ^{125}I-α-bungarotoxin binding to membranes containing *Torpedo* $\alpha 1$ receptors with an IC_{50} of 100 nM. These data suggest that radiolabeled EB may be a useful ligand for studying several nicotinic receptor subtypes. Supported by NIH NS22472.

Section 6: Receptor regulation

P 36   [³H]CYTISINE BINDING IN RAT PRIMARY NEURONAL CELLS IN CULTURE: A MODEL SYSTEM TO STUDY NEURONAL NICOTINIC ACETYLCHOLINE RECEPTORS. M. I. Dávila-García, S. S. Oasba, R. A. Houghtling and K. J. Kellar. Department of Pharmacology, Georgetown University School of Medicine, Washington, DC 20007.

The study of factors that regulate neuronal nicotinic acetylcholine receptors has been impeded by the lack of a suitable *in vitro* system. Thus, we have used a primary neuronal cell culture system from embryonic day 18 rat fetal brain stem (including the thalamus, hypothalamus, midbrain and pons) and have characterized [³H]cytisine binding to nicotinic acetylcholine receptors present in these primary neurons. Our results demonstrate that after 15 days in culture [³H]cytisine binding to brain stem neurons is saturable and specific, with a Kd of 0.7 nM and a Bmax of ≈ 50 fmol/mg protein. [³H]Cytisine binding was competitively displaced by nicotine with an IC_{50} of 20 nM and by carbachol and dihydro-ß-erythroidine with IC_{50} values of <1 uM. These studies indicate that this cell culture system may be a suitable model in which to elucidate the properties of neuronal nicotinic acetylcholine receptors, to define the subunit composition of the receptors labeled by high affinity nicotinic ligands and to study the mechanisms involved in the regulation of these receptors. This work was supported by NIH grants DA05417 and DA06486.

P 37   IDENTIFICATION AND CHARACTERIZATION OF α-BUNGAROTOXIN SENSITIVE NICOTINIC RECEPTORS IN IMMORTALIZED HIPPOCAMPAL NEURONS J. Komourian and M. Quik. Dept. of Pharmacology, McGill University, Montreal, Canada H3G 1Y6.

The functional significance of neuronal nicotinic α-bungarotoxin (α-BGT) receptors is at present not fully understood. As an approach to study the role and regulation of α-BGT sites in the CNS, we have selected an immortalized hippocampal cell line to serve as a model. The H19-7 cell line (courtesy of Dr. Wainer, University of Chicago) was produced by transforming embryonic day 18 rat hippocampal cells with a temperature sensitive mutant of the simian virus 40 large tumor antigen. These immortalized cells proliferate at the permissive temperature of 33°C (high serum medium), but cease to divide and commence to differentiate into neuronal cells at the nonpermissive temperature of 39°C (low serum medium containing differentiating agents). ¹²⁵I-α-BGT binding studies showed that undifferentiated cells displayed very few α-BGT sites in contrast to the differentiated state. Saturation analysis using differentiated cells yielded a K_d of 1.30 ± 0.05 nM (n=3) and a B_{max} of 11.2 ± 1.73 fmoles/10⁵ cells (n=3). A developmental profile demonstrated that receptor expression was maximal 7 days after addition of the differentiation medium. ¹²⁵I-α-BGT competition binding studies to the H19-7 cell membranes showed that nicotinic but not muscarinic receptor ligands interacted at the α-BGT site, with IC_{50} values as follows: α-BGT, 56 ± 3 nM (n=4); d-tubocurarine, 6.3 ± 0.4 μM (n=3); nicotine, 100 ± 6 μM (n=3); carbachol, 90 μM (n=2) and DMPP, 90 ± 4 μM (n=3). Thus the present studies demonstrate that H19-7 cells express a nicotinic α-BGT receptor. Work will now be initiated to examine the potential role of this nicotinic receptor population in neuronal tissue.

P 38   CHRACTERISTICS OF [³H]CYTISINE AND (-)-[³H]NICOTINE BINDING TO THE MEMBRANE PREPARATIONS OF MOUSE FIBROBLAST M10 CELLS STABLY EXPRESSING α4β2 NICOTINIC ACETYLCHOLINE RECEPTORS. X. Zhang, Z-H. Gong and Agneta Nordberg Department of Clinical Neurosciences and Family Medicine, Division of Nicotine Research, Karolinska Institute, Huddinge Universtiy Hospital, S-141 86, Huddinge, Sweden

This study examined the binding characteristics of [³H]cytisine and (-)-[³H]nicotine in the membrane preparations of the mouse fibroblast M10 cells which stably express α4β2 nicotinic acetylcholine receptors (nAChRs), a major subtype of neuronal nAChRs. The kinetic studies revealed that the associated constants were 0.047 min⁻¹nM⁻¹ and 0.011 min⁻¹nM⁻¹ for [³H]cytisine and (-)-[³H]nicotine, respectively. The K_D value (0.37 nM) for [³H]cytisine revealed by saturation binding assays was about 10-fold lower than that (4.99 nM) for (-)-[³H]nicotine. The Bmaxs for [³H]cytisine and (-)-[³H]nicotine are the same. Both [³H]cytisine and (-)-[³H]nicotine binding was stereoselectively displaced from one high affinity site by the enantiomers of nicotine. The affinities of the enantiomers of nornicotine for the α4β2 nAChRs are 10-fold higher compared with those observed in rat brain homogenates. Interestingly all of the agonists studied showed an one-binding site model while the antagonists (+)-tubocurarine and dihydro-β-erythroidine showed a two-binding site model in competing both [³H]cytisine and (-)-[³H]nicotine binding. (-)-Nicotine treatment significantly up-regulated the number of (-)-[³H]nicotine binding sites. The ED_{50} value of (-)-nicotine in up-regulating the number of (-)-[³H]nicotine binding sites was about 1000-fold higher than its Ki value in displacing (-)-[³H]nicotine binding in the membrane preparations of the M10 cells. These results indicate that the M10 cells can be a useful system for study of the regulation of nAChRs.

Modulation of Human Neuronal Nicotinic Acetylcholine Receptor (nAChR) Activation by L-Type Calcium Channel Antagonists Diana L. Donnelly-Roberts, Murali Gopalakrishnan, Stephen P. Arneric and James P. Sullivan Neuroscience Research (D-47W), Abbott Laboratories, Abbott Park, IL, 60064-3500

Evidence is rapidly emerging to indicate that nAChRs may be modulated through sites distinct from the classical acetylcholine (ACh) binding sites (Arneric et al, Psychopharmacology:4th Gen. 1994, in press). For example, recent studies suggest that the neuronal nAChRs present on chromaffin cells contain a dihydropyridine-sensitive site whose occupation at clinically relevant (low micromolar) concentrations blocks membrane depolarization (Lopez et al., Eur. J. Pharmacol., 1993 247:199-207). The subunit specificity, if any, of this modulatory site is not known. A recent study (Lukas J. Pharmacol. Exp. Ther. 1993,265:292-302] has suggested that (-)-nicotine-induced cation (^{86}Rb+) efflux in the human neuroblastoma cell line IMR 32 reflects activation of the $\alpha 3 \beta 4$ subtype. In the present study the ability of a number of calcium channel antagonists to regulate the activation of the nAChRs present in IMR 32 cells was examined. Several dihydropyridine L-type Ca++ channel blockers (nimodipine, PN 200110, nifedipine and (R)+-BAY K 8644) inhibited (-)-nicotine-induced cation efflux in a dose-dependent manner with IC$_{50}$ values ranging from 1-10 μM. Two non-dihydropyridine L-type calcium channel blockers, verapamil and diltiazem also inhibited nAChR activation exhibiting IC$_{50}$ values of 0.7 μM and 0.6 μM, respectively. In contrast, w-Conotoxin GVIA, an N-type Ca++ channel blocker, and w-Agatoxin IVA, a P-type Ca++ channel blocker had no effect on nAChR activation at concentrations up to 10 μM. Further, the non-selective channel blocker, cadmium, had no effect on modulatory role of the L-type Ca++ channel antagonists suggesting that the effects of these agents on nAChRs are not mediated via an interaction with calcium channels. Thus, the results of the present study suggest that activation of the $\alpha 3 \beta 4$ nAChR subtype can be modulated by L-type, but not N- or P-type Ca++ channel antagonists. Whether this effect is selective to this nAChR subtype remains to be determined.

ON THE FALL AND RISE OF NEURONAL ALPHA-BUNGAROTOXIN BINDING PROTEINS INDUCED BY HETEROLOGOUS SENSE OR ANTISENSE ALPHA7 SUBUNIT TRANSGENE EXPRESSION IN HUMAN SH-SY5Y NEUROBLASTOMA CELLS. Elzbieta Puchacz, Cynthia Eisenhour and Ronald J. Lukas. Division of Neurobiology, Barrow Neurological Institute, Phoenix, AZ 85013 USA.

Like their non-neoplastic neuronal analogs of neural crest-origin, cells of the SH-SY5Y human neuroblastoma express two types of pharmacologically and structurally-distinct nicotinic acetylcholine receptors (nAChR) and at least five nAChR subunits: alpha3, alpha5, alpha7, beta2, and beta4. Ganglia-type nAChR in these cells, which mediate functional ion flux responses that are insensitive to alpha-bungarotoxin (Bgt), are distinct from neuronal/nicotinic Bgt binding sites (nBgtS), which bind ^{125}I-labeled Bgt (I-Bgt) with high affinity. However, relationships between expressed subunit gene products and these two nAChR subtypes are not established.

In investigate this issue, we have examined I-Bgt binding and nAChR function in SH-SY5Y cells engineered to overexpress rat alpha7 transgenes in sense or antisense orientations. Transfected cells can actively transcribe the transgenes as shown by Northern blot analyses. Cells harboring sense-strand vectors express 10 times higher specific I-Bgt binding than control cells, whereas a 75% reduction has been seen in one study of antisense alpha7 expression. In neither case is there an effect on nAChR function or agonist binding. Cyclosporin treatment of sense alpha7-expressing cells reduces surface I-Bgt binding by 40-50% whereas I-Bgt binding in control cells expressing only native nBgtS is unaffected or increased slightly. These results suggest that human alpha7 subunits contribute to expression of native nBgtS (but not ganglia-type nAChR) in SH-SY5Y cells but that such expression is cyclosporin- insensitive, in contrast to the cyclosporin-sensitive expression of exogenous rat alpha7-containing I-Bgt binding sites, which may assemble as cylcophilin-dependent and cyclosporin-sensitive homooligomers.

EPSILON SUBUNITS RESCUE MUSCLE NICOTINIC ACETYLCHOLINE RECEPTOR EXPRESSION IN BUTYRATE-TREATED BC$_3$H-1 CELLS. Elzbieta Puchacz, Linda Lucero and Ronald J. Lukas. Division of Neurobiology, Barrow Neurological Institute, Phoenix AZ 85013 USA.

Mammalian muscle nicotinic acetylcholine receptors (nAChR) are heterooligomeric membrane proteins composed of four types of subunits: alpha, beta, delta and either gamma or epsilon. Adult muscle nAChRs contain the epsilon subunit, and the switch from fetal gamma to adult epsilon subunit expression accounts for a functionally- and developmentally-relevant switch in muscle nAChR phenotype about the time of motor neuronal innervation of muscle in vivo. However, mechanisms that underlie these transitions haven't been elucidated. In studies testing the effects of different agents that are known to alter second messenger signaling or gene transcription, we found that treatment of cells of the BC$_3$H-1 muscle cell line with sodium butyrate mimics at least some of the changes in nAChR that occur developmentally: (1) numbers of cell surface and total membrane nAChR decline >60%, (2) nAChR function declines >90%, (3) levels of nAChR alpha, beta and delta mRNAs are reduced severalfold, and (4) there is a complete loss of fetal nAChR gamma subunit-encoding transcripts.

In this study we have tested the hypothesis that the loss of nAChR function in butyrate-treated BC$_3$H-1 cells due to suppression of gamma subunit gene expression can be rescued by expressing exogenous, transgenic epsilon subunits. We have now found that expression of epsilon transgenes from sense orientation constructs, but not from antisense orientation constructs, under CMV promoter control retain 15-20% of control cell function in instances where gamma subunit expression is abolished by butyrate treatment. Hence, this drug treatment/transgene expression system is suitable for studies of adult nAChR pharmacology and function and of native or mutated nAChR epsilon or gamma transgene expression in mammalian muscle-like cells.

P 42

IS THE NICOTINIC RECEPTOR UPREGULATION FOLLOWING CHRONIC NICOTINE TREATMENT A RESULT OF STIMULATION-INDUCED RECEPTOR INACTIVATION?

Peter P. Rowell, Department of Pharmacology and Toxicology, University of Louisville School of Medicine, Louisville, Kentucky 40292, USA.

Many studies have found that twice-daily injections of nicotine to animals results in an increase in the density of nicotinic receptors in most brain areas. It has been hypothesized that this nAChR upregulation is a consequence of long-term receptor desensitization. Recently, several studies have indicated that the continuous administration of nicotine is less effective at producing nAChR upregulation than twice-daily treatment. In this study, the same doses of nicotine (1.2 and 2.4 mg/kg/day) were administered to rats either 2,4,8 x/day or continuously, and the nicotine blood and brain levels and nAChR density was examined in several brain areas and compared with the concentrations of nicotine necessary to produce nAChR desensitization and stimulation *in vitro*. It was found that continuous nicotine infusion maintained desensitizing brain nicotine concentrations at all times but was less effective at producing receptor upregulation than intermittent administration which, in contrast to continuous administration, reached stimulating nicotine concentrations but also fell below desensitizing concentrations during the day. We suggested that, in addition to the short-lived desensitization, stimulating concentrations of nicotine cause long-term receptor inactivation, a process that could be maintained for many hours between doses. We propose that this stimulation-induced receptor inactivation is responsible for the nAChR upregulation seen with intermittent nicotine administration whereas the continuous administration of the same daily dose of nicotine, while achieving steady-state brain concentrations more than sufficient to produce complete receptor desensitization, may not cause nAChR upregulation unless a dose high enough to produce receptor inactivation is administered. If correct, this proposal has important implications with respect to the delivery of nicotine by cigarette smoking versus the transdermal patch.

P 43

NICOTINE UP-REGULATES HIGH AFFINITY $\alpha4\beta2$ NICOTINIC ACETYLCHOLINE RECEPTORS THROUGH POST-TRANSLATIONAL MECHANISMS.

Merouane Bencherif, Kathy Fowler, Ronald J. Lukas[#], and Patrick M. Lippiello. R & D Dept., R. J. Reynolds Tobacco Company, Winston-Salem, NC 27102, and # Barrow Neurological Institute. St. Joseph's Hospital and Medical Center, Phoenix, AZ 85013.

There is a consensus that the number of high affinity $[^3H]$-L-nicotine binding sites (HnAChR) increases following chronic nicotine exposure but the mechanisms have not yet been elucidated. We have used a combination of primary cultures of fetal rat brain cortex and clonal cell cultures of fibroblasts stably transfected to express the predominant brain subtype. These stably transfected cells which express alpha4 and beta2 subunits genes under dexamethasone-inducible promoter (M10 cell line from Whiting et al., 1991) were used to evaluate the contribution of transcriptional, translational and post-translational mechanisms in HnAChR regulation. Chronic nicotine exposure resulted in a marked increase in HnAChR binding site density in both primary fetal rat brain cultures and M10 cells reaching $260 \pm 40\%$ of control values (n=11). Up-regulation of HnAChR in M10 cells was evidenced at concentrations as low as 10nM and reached maximal effect between 1 and 10uM. Scatchard analyses of binding data indicate a change in Bmax with no change in affinity. Nicotine alone had no effect on the dexamethasone-inducible promoter. Upon removal of dexamethasone, both control and up-regulated HnAchR decreased with a single rate constant (k = -0.69; $t_{1/2}$ = 25 hours). When nicotine was applied at the time of dexamethasone removal, HnAChR levels were maintained at approximately 100% (vs. 50% for non-treated cells; EC_{50} = 100 nM) in control but not in nicotine-pretreated cells. Northern blot analyses on fetal rat brain indicate no significant change in $\alpha4$ and $\beta2$ mRNA levels between control and nicotine-treated cells. These results suggest that nicotine enhances the half life of a pool of receptors convertible into high-affinity conformers but also suggest the existence of an additional pool of low affinity non-convertible receptors. A model for the mechanisms of up regulation of the $\alpha4\beta2$ HnAChR following chronic exposure to nicotine is presented and suggests that receptor occupancy can trigger postranslational modifications, resulting in increased half-life of the high affinity conformer.

P 44

CHRONIC NICOTINE EXPOSURE DECREASES THE ACTIVATION OF $\alpha4\beta2$ BUT NOT $\alpha3\beta2$ NEURONAL NICOTINIC RECEPTORS EXPRESSED IN *XENOPUS* OOCYTES.

Y.N. Hsu, J. Amin, D. Weiss and L. Wecker. Departments of Pharmacology and Therapeutics and Physiology and Biophysics, Univ. South Florida College of Medicine, Tampa, FL 33612.

Recent studies in our laboratory have shown that the chronic administration of nicotine increases nicotine-mediated $[^3H]DA$ release, but decreases nicotine-mediated $[^3H]ACh$ release from brain slices. Although the mechanism(s) mediating these effects are not well understood, evidence suggests that chronic nicotine exposure may differentially modulate the $\alpha3\beta2$ receptor subtype mediating the induced release of DA and the $\alpha4\beta2$ receptor subtype mediating the induced release of ACh. To test this possibility, cRNAs for $\alpha3$ and $\beta2$ or $\alpha4$ and $\beta2$ subunits were co-injected into *Xenopus* oocytes and nicotine-activated currents were measured prior to and after 24 hrs incubation with nicotine using the two-electrode voltage-clamp technique. Currents in oocytes incubated with control media increased over 24 hrs due to increased receptor expression. Nicotine-activated currents in oocytes expressing $\alpha3\beta2$ receptors were not affected significantly by 24 hr incubation with nicotine; currents increased 6.8-, 5.9- and 5.2-fold following incubation with 0, 20 and 100 nM nicotine, respectively. In contrast, nicotine-activated currents in oocytes expressing $\alpha4\beta2$ receptors decreased in response to incubation with nicotine; current in control oocytes increased 5.8-fold, whereas current in oocytes incubated with 20 nM nicotine increased only 2.2-fold and current in oocytes incubated with 100 nM actually decreased. Because this reduced response could be reversed by incubating oocytes for 90 min in control medium, results suggest that the $\alpha4\beta2$ receptors were expressed, but desensitized. These findings suggest that the differential effect of chronic nicotine administration on the release of DA and ACh may be due, at least in part, to differences in the desensitization of the two receptor subtypes mediating the release of these neurotransmitters. (Supported by grants #0411 from the STRC, Inc. and #AA09212 from the NIH.)

ON THE ROLE OF ACETYLCHOLINESTERASE AND NICOTINIC ACETYLCHOLINE
RECEPTORS IN DENERVATION SUPERSENSITIVITY IN THE FROG CARDIAC
GANGLION. P.B. Sargent, E.N. Garrett, H.L. Wilson, and S.D. Matthews. Depts. of
Stomatology and Physiology and the Neurosciences Graduate Program, UCSF, San
Francisco, CA 94143.

Chronic denervation of parasympathetic neurons in the frog cardiac ganglion leads to an
increase in their sensitivity to acetylcholine but does not alter sensitivity to carbamylcholine
(J. Physiol., 445:249-260, 1992). This suggests that denervation supersensitivity results from
a reduced effectiveness of acetylcholinesterase (AChE). To learn whether denervation leads
to an outright loss of AChE, we measured AChE in extracts of sham-operated and denervated
ganglia. Denervation resulted in a significant *increase* in the V_{max} for total AChE. Previous
work (J. Neurobiol. 21: 938-949, 1990) suggested that this increase is attributable to enhancement of an intracellular pool of enzyme. When extracellular AChE was measured specifically,
it was found not to be altered significantly by denervation. Quantitative histochemical studies
further showed no denervation-induced change in AChE activity at the surface of ganglion
cells. These studies suggest that denervation supersensitivity is not caused by a loss of
AChE. Nor is it apparently explained by an increase in the number of cell surface nicotinic
acetylcholine receptors (AChRs), measured as the number of binding sites for ^{125}I-neuronal-
bungarotoxin (κ-bungarotoxin) or for ^{125}I-mAb 22, a cross-reacting anti-electric organ AChR
antibody (gift of Dr. Jon Lindstrom, UPenn). Finally, denervation does not appear to alter the
distribution of AChRs relative to AChE, thus making it unlikely that denervation supersensitivity to ACh results from the migration of AChRs to areas of the cell surface containing
relatively little AChE. (Supported by NIH grant NS 24207.)

REGULATION OF NICOTINIC RECEPTORS IN RAT BRAIN FOLLOWING QUASI-IRREVERSIBLE
NICOTINIC BLOCKADE BY CHLORISONDAMINE AND CHRONIC TREATMENT WITH NICOTINE H. El-
Bizri and P. B. S. Clarke. Dept. of Pharmacology and Therapeutics, McGill
University, Montreal, Canada H3G 1Y6

Chronic administration of nicotinic agonists in vivo increases the density of
brain nicotinic binding sites. We tested the hypothesis that this up-regulation
results from agonist-induced inactivation of nicotinic receptors. ^3H-nicotine
and ^{125}I-α-bungarotoxin (^{125}I-αBTX) binding were examined following in vivo
treatment with the chronic CNS nicotinic blocker chlorisondamine(10 mg kg^{-1} SC),
given either alone or in combination with chronic nicotine administration (1 mg
kg^{-1} SC b.i.d. for 12 days). In the absence of chlorisondamine pretreatment,
chronic nicotine administration increased the B_{max} of ^3H-nicotine binding in the
cerebral cortex, striatum, midbrain and hippocampus; K_D was unchanged. This upregulation was neither mimicked nor blocked by chlorisondamine pretreatment,
despite persistent blockade of acute locomotor responses to nicotine. Chronic
nicotine treatment also increased the B_{max} (but not K_D) of ^{125}I-α-BTX binding in
cerebral cortex, hippocampus and midbrain. Chlorisondamine altered neither Bmax
nor K_D when given alone, but significantly attenuated the nicotine-induced upregulation of toxin binding sites in midbrain, with a similar trend in the other
two regions. These findings suggest that in the CNS, up-regulation of ^3H-
nicotine binding sites is not determined by the functional status of the nicotinic
receptors. In contrast, up-regulation of ^{125}I-α-BTX binding sites may result
from chronic receptor *activation*.

PERSISTENT PRESENCE OF TRITIUM IN RAT BRAIN FOLLOWING IN VIVO ADMINISTRATION OF
^3H-CHLORISONDAMINE: POSSIBLE INTRACELLULAR ACCUMULATION H. El-Bizri and P.B.S.
Clarke. Dept. of Pharmacology and Therapeutics, McGill University, Montreal,
Canada H3G 1Y6.

Chlorisondamine (CHL) is a bisquaternary ganglion blocker which, after a single
administration, produces a central nicotinic blockade that can last at least 12
weeks. The aim of the present study was to test whether CHL (or a metabolite)
persists in the brain during this central nicotinic blockade, and to determine its
regional and subcellular distribution. Rats were injected with ^3H-CHL (10 μg
icv) and were allocated to five groups (1 day or 1, 3, 9, or 12 weeks survival; n
= 7 per group). In each group, four rats provided coronal brain sections which
were processed for dry film autoradiography; in the remaining subjects, tritium
was measured in subcellular fractions prepared from various brain regions. At
all survival times, radioactivity was particularly dense in the substantia nigra,
dorsal raphé, and in the cerebellum. Radioactivity was present in membrane
fractions but was virtually absent from soluble fractions. Tritium remained
bound to the synaptosomal fraction (P2) even when synaptosomes were lysed with
water. To test for retrograde transport (and hence intracellular accumulation),
0.5 μg ^3H-CHL was injected into the striatum. Seven days later, the ipsilateral
substantia nigra pars compacta was heavily labelled. These findings suggest that
substantial amounts of chlorisondamine (or a metabolite) persist for many weeks in
the brain, become tightly bound to membranes, and can become concentrated in
neuronal cell bodies.

P 48   RFLP ANALYSIS OF THE RELATIONSHIP BETWEEN THE INHERITANCE OF STRAIN-SPECIFIC nACHR α7 ALLELES, NICOTINE-INDUCED SEIZURE SENSITIVITY AND LEVELS OF [^{125}I]-α-BUNGAROTOXIN BINDING J.A. Stitzel, D.A. Farnham, and A.C. Collins, IBG, Univ. of CO. Boulder, CO. 80309.

Previously, we have demonstrated that mouse strains vary in their sensitivity towards nicotine-induced seizures. Results of classic genetic cross studies using two strains (C3H/2ibg and DBA/2ibg) that differ maximally for this measure indicate that the inheritance of seizure sensitivity fits a single gene, two allele model. In addition, seizure sensitivity has been shown to display a significant correlation with the level of hippocampal [^{125}I]-α-bungarotoxin (α-BTX) binding such that those animals having higher levels of binding are more sensitive to nicotine-induced seizures. Taken together, these findings suggest that the gene which encodes the α-BTX-binding receptor subunit might be critically involved in determining strain-specific differences in sensitivity toward nicotine-induced seizures. Examination of C3H and DBA genomic DNA for restriction fragment length polymorphisms (RFLPs) established that these strains are polymorphic at the nACHR α7 gene which codes for a brain α-BTX binding nACHR subunit. Therefore, F1 x DBA, F1 x C3H and F2 animals derived from genetic crosses between C3H and DBA mice were compared for seizure sensitivity, α-BTX binding and α7 genotype. Unlike the significant correlation observed between seizure sensitivity and the number of hippocampal α-BTX binding sites, no significant correlation between α7 genotype and seizure sensitivity was found. Moreover, no detectable relationship between α7 genotype and the number of α-BTX binding sites was observed. These findings suggest that strain-specific differences in some gene or genes other than that which codes for the α-BTX binding nACHR subunit is critical in determining strain-specific differences in nicotine induced seizure sensitivity and the level of hippocampal α-BTX binding. (This work supported by DA05131, DA00197, and AA07464).

P 49   NEURONAL NICOTINIC ACETYLCHOLINE RECEPTORS IN SCHIZOPHRENIA, Leonard, S., Adler, L.E., Bickford, P.C., Hall, M., Rollins, Y., Breese, C., Logel, J., Drebing, C., Byerley, W., Coon, H., Freedman, R., U. of Colorado and Vet. Admin. Medical Centers, Denver, CO.

Schizophrenic individuals show a deficit in their ability to process sensory information, characterized by an abnormal electrophysiological response to auditory stimuli, a failure to "gate" or filter out the second of two paired clicks. Nicotine has been shown to transiently normalize the deficit in schizophrenics as well as in an animal model of this habituation, suggesting a critical role for cholinergic innervation to the hippocampus from the medial septal nucleus. In the animal model, the auditory gating deficit can be reproduced with agents which block a specific subset of the nicotinic receptor family, the α-bungarotoxin binding receptor, α7. We have isolated two cDNA clones covering the full length of the coding region, from a human postmortem brain library. A northern blot indicates that α7 transcripts in human brain are of different sizes than those found in either rat or neuroblastoma cells, and are more highly expressed in cingulate than hippocampus. Using autoradiographic analysis with [^{125}I]-α-bungarotoxin, we found that the expression of the α7 receptor in hippocampus dissected postmortem from schizophrenics appears to be decreased compared to control subjects. An independent full genome scan for linkage of both the gating deficit and schizophrenia resulted in a positive LOD score for markers around the chromosomal locus of the human α7 gene, 15q14. These results suggest that decreased or aberrant expression of the α7 neuronal nicotinic receptor may be closely associated with a failure to process sensory information in schizophrenia.

P 50   ADAPTIVE MECHANISMS ASSOCIATED WITH CHRONIC NICOTINE TOLERANCE AND WITHDRAWAL IN BOVINE ADRENAL CHROMAFFIN CELLS. A.E. Bullock and A.S. Schneider. Dept. Pharmacology & Toxicology, Albany Medical College, Albany, NY 12208

We investigated cellular mechanisms of adaptation to chronic low-dose nicotine exposure in cultured adrenal chromaffin cells. Chromaffin cells take up calcium and release catecholamines upon exposure to nicotine. Preexposure to nicotine results in tolerance and a depression of these responses upon subsequent nicotine challenge. Following chronic preexposure to 1.0 µM nicotine for 5-7 days and a short period of nicotine withdrawal, chromaffin cells exhibited an increased rate of desensitization of nicotinic receptor-mediated catecholamine release relative to control cells not previously exposed to nicotine. Chronic preexposure to 1 uM nicotine for 5-7 days and withdrawal (3 hr) also resulted in a change in the shape of the nicotine dose-response curve and an over-shoot in catecholamine release stimulated by 10 µM nicotine. The number of specific binding sites for mAb 35, a monoclonal antibody to the nicotinic receptor, was shown to increase following 5 day chronic nicotine preexposure and withdrawal. The rank order of nicotinic agonist potency for stimulation of catecholamine release was determined in order to obtain information on the identity of the chromaffin cell nicotinic receptor subtype. Agonist potency and mAb 35 binding would be consistent with more than one type of alpha subunit (α3 and α5) or possibly more than one nicotinic receptor subtype on chromaffin cells. Cellular mechanisms of chronic nicotine tolerance and dependence may involve several cellular processes including up-regulation of nicotinic receptors, a hypersensitive catecholamine release response following nicotine withdrawal and an increase in rate of receptor desensitization. Supported by grants NIH BRSG SO7RR05394-31 and American Heart Assoc NY State Affil. 90-49G.

NICOTINIC RECEPTOR SUBTYPES CONTROLLING THE SECRETAGOGUE AND MITOGENIC
EFFECTS OF NICOTINIC AGONISTS IN SMALL-CELL LUNG CARCINOMA CELL LINES
A. Codignola, P. Tarroni, M.G. Cattaneo, L.M. Vicentini, F. Clementi & E. Sher. CNR Center of
Cytopharmacology and Dept. of Medical Pharmacology, University of Milan, Milan, ITALY.
We have recently described the presence of neuronal type nicotinic receptors of an α-Bungarotoxin
(αBgtx) binding sites in human small cell lung carcinoma (SCLC) cell lines (B. Chini et al., 1992; P.
Tarroni et al., 1992). These receptors mediate both the secretagogue and mitogenic effects of nicotine in
these cancer cells (M.G. Cattaneo et al., 1993). We now found that different related polypeptide
nicotinic antagonists (αBgtx, α-Conotoxin MI (αCtx) and κBungarotoxin (κBgtx)) are all potent blockers
of both the secretagogue and mitogenic effects of nicotinic agonists in SCLC cells. Nicotine and
Cytisine stimulate [^{3}H]Serotonin ([^{3}H]5HT) release from three different SCLC cell lines (NCI-H-69,
NCI-N592 and GLC8) in a time- and dose-dependent manner, the EC_{50} being 20 ± 2, 22.5 ± 1 and
23.5 ± 1(nicotine) and for the two drugs, respectively. Nicotine-induced [^{3}H]5HT release was
completely dependent on the presence of external Ca^{2+}, suggesting that Ca^{2+} influx is a crucial step in
this phenomenom. αBgtx, αCtx and κBgtx antagonized nicotine- and cytisine-induced [^{3}H]5HT release
in a dose-dependent manner (IC_{50s} of 1 nM, 10 pM and 1 pM, respectively).
Nicotine- and cytisine-stimulated SCLC cell proliferation was also completely prevented by αBgtx 1.53
μM and αCtx 1.1 μM, while the mitogenic effects of serotonin (M.G.Cattaneo et al., 1993) were not
affected. Besides the α_3, α_5 and β_2 nicotinic subunits we have already described in SCLC cells, we have
now found evidence, by PCR analysis, for the presence in SCLC cells of the α_7 and β_4 nicotinic
subunits. β_4 is known to confer cytisine sensitivity to neuronal-type nicotinic receptors, and α_7 is known
to code for the most abudant αBgtx-sensitive nicotinic receptor in the nervous system.
Our results suggest that the αBgtx-sensitive nicotinic receptors of SCLC cells, possibly coded by the α_7
subunit, play a crucial role in mediating the biological effects of nicotine in this very aggressive lung
cancer.

Section   7:   Effects on gene expression

NICOTINE INDUCES FOS INTENSELY IN THE PARVOCELLULAR
PARAVENTRICULAR NUCLEUS, THE PREOPTIC AND THE LATERAL
HYPOTHALAMUS IN RATS. Bryan Bienvenu, Hideo Kiba, Jayashree Rao* and
A. Jayaraman. Depts. of Pediatrics* & Neurology, LSU School of Medicine, New
Orleans, LA, 70112, U.S.A.
Subcutaneous injections of nicotine (0.4 to 1.0 mg/kg) in adult male rats resulted in the
induction of c-fos gene most prominently and selectively in the parvocellular, but not the
magnocellular subdivision of the hypothalamic paraventricular nucleus. A significant
number of neurons in the superficial regions of the supraoptic nucleus also contained
Fos reactivity, but only with larger doses (1.0 mg/kg) of nicotine. Fos immunoreactive
neurons were also prominent in the supramammillary regions. The lateral preoptic area,
the anterior and posterior aspects of the lateral hypothalamus contained significant
number of Fos immunoreactive neurons. The medial preoptic area, the suprachiasmatic
and the periventricular nuclei of the hypothalamus were relatively free of Fos reactive
neurons. Injections of mecamylamine completely abolished nicotine induced Fos
immunoreactivity in all of these cases. These results suggest that acute injections of
nicotine induce intense Fos expression in two major areas of the rat hypothalamus,
namely 1. the CRF neurons of the parvocellular paraventricular nucleus, an area of the
hypothalamus recognized to mediate stress response and 2. the lateral preoptic areas
and the lateral hypothalamus, regions of the hypothalamus strongly implicated in
intracranial self-stimulation behavior. Supported by The Smokeless Tobacco Research
Council, N.Y., and the Department of Defense, U.S.A.

P 53   EXPRESSION OF NEURONAL FOS PROTEIN AFTER REPEATED ADMINISTRATION OF NICOTINE IN THE RAT BRAIN. O.Salminen and L.Ahtee. Dept. of Pharmacy, Div. of Pharmacology & Toxicology, University of Helsinki, Helsinki, Finland.

Previous studies have shown increases in Fos immunostaining (IS) in various rat brain areas after acute nicotine (1,2). High levels of Fos IS were seen in superior colliculus (SuG), the medial terminal nucleus of the accessory optic tract (MT), interpeduncular nucleus (IP), supraoptic nucleus (SO) and paraventricular hypothalamic nucleus (PaV) after a single nicotine injection (1 or 2 mg/kg, s.c.). To study the effects of repeated administration, nicotine (1 mg/kg, s.c.) was given to rats 2 times daily for 5 days. Rats were deeply anesthetized with pentobarbital one hour after last injection and perfused with 0.9% phosphate-buffered saline followed by 4% paraformaldehyde in phosphate buffer pH 7.4. Sections (40 μm) were cut on a cryostat and processed with avidin-biotin immunohistochemical method using rabbit anti c-fos antibody "DCH1" 1:4000. One group of rats was perfused 24 hours after last nicotine injection. Repeated administration of nicotine induced Fos IS similar to what was seen after a single acute nicotine injection. Most of these areas express high densities of nicotine binding with in vitro autoradiography. Similarly to acute nicotine (1) repeated nicotine failed to induce Fos IS in areas rich of dopaminergic innervation such as striatum, substantia nigra and VTA . Repeated administration did not cause any noticeable tolerance in Fos IS. Furthermore, 24 hours after cessation of repeated administration, no Fos IS was seen in the above mentioned areas. Thus, this procedure of nicotine administration was not sufficient to cause any long-lasting changes in expression of Fos protein, which would have indicated nicotine dependence.
1. Ren T and Sagar SM: Brain Res Bull, 1992, 29:589-597.
2. Korhonen S, Salminen O, Ahtee L, XLVI Annual Meeting of the Finnish Pharmacological Society, 1994.

P 54   CLASSICAL CONDITIONING OF cFOS PROTEIN EXPRESSION FOLLOWING EXPOSURE TO NICOTINE- OR COCAINE-PAIRED ENVIRONMENTS: ROLE OF FOREBRAIN STRUCTURES. J. McCoy, S. Matta, J. Valentine, B. Sharp. Endocrine-Neurosci. Labs, Minneapolis Medical Research Fndn, Depts. of Medicine, Hennepin County Medical Center and the University of Minnesota 55404
In human clinical trials, craving can be evoked by the presentation of contextual stimuli previously associated with the effects of nicotine (Nic) or cocaine (Coc). These salient environmental cues can therefore acquire secondary reinforcing properties. Since c-fos gene expression, a marker of neuronal activation, is induced by both Nic and Coc, this study used cFos protein immunocytochemistry (ICC) to determine if environmental stimuli previously paired with Nic or Coc could activate cFos in forebrain structures. Male rats were acclimated and randomly assigned to one of five groups: 1) Nic (0.5 mg/kg; ip)/conditioned (Nic/Cond) 2) Nic pseudoconditioned (Nic/Pseudo) 3) Coc (10 mg/kg; ip)/conditioned (Coc/Cond) 4) Coc/pseudoconditioned (Coc/Pseudo) 5) vehicle control (Veh). For 10 d of conditioning, rats were injected with Nic or Coc, placed in a conditioning chamber for 60 min, returned to their home cages, then injected with saline 4 hr later. For pseudoconditioning, rats received saline in the salient environment, then drug in their home cages 4 hr later. Control rats received saline in both environments. 48 hours after the final training session, every rat was injected with saline, placed in the conditioning chamber for 60 min, transcardially perfused and brains were cryosectioned for ICC. Minimal cFos ICC was observed in the Veh, as well as both Pseudo groups. cFos activation in response to conditioning with Nic or Coc was seen to a similar degree in the cingulate gyrus, claustrum and cerebral cortex. Significantly greater cFos ICC was found in the following regions from Nic/Cond rats compared to Coc/Cond: amygdala; pyriform cortex, which was several fold greater; and caudate/putamen which did not respond in the Coc/Cond. These results suggest that forebrain structures may function in the development and/or expression of environment-specific appetitive conditioning to drugs of abuse and that a subset of these forebrain regions are either more responsive or specific to nicotine-conditioning. (NIH DA-03977)

P 55   NICOTINE INDUCES CHANGES IN CNS GENE EXPRESSION. R.L. Martone, J.L. Jeffries-Griffor, S.P. Williams, & S.L. Orr Dept. of Molecular Genetics and Protein Chemistry, Pfizer Inc., Central Research Division, Groton, CT 06340 USA

The addictive characteristic of nicotine is believed to be mediated by the effects of this drug on the mesolimbic dopaminergic system. We have employed the differential display method of Liang and Pardee (Science, 257:967-971;1992) to identify nicotine-induced changes in gene expression in one component of the mesolimbic system, the ventral tegmental area (VTA). RNA was isolated from pooled VTA tissue of 30 rats which were subjected to daily injections of nicotine tartrate (1.5 mg/kg) for 2 weeks, and from pooled VTA tissue of 30 saline injected control rats. Samples were reverse transcribed and PCR amplified using 12 different oligo dT primers paired with 10 arbitrary 10mer primers. Complementary DNA bands corresponding to genes that appear to be differentially expressed in response to nicotine were isolated and sequenced. Most of the 97 cDNA bands isolated in this manner consist of novel sequences, but include several identifiable sequences such as glutamate receptor and several mitochondrial genes, including ATP synthase, and mitochondrially encoded cytochrome oxidase subunits. Efforts are underway to clone and further characterize the novel sequences that appear to be regulated by nicotine.

EFFECT OF NICOTINE ON EXPRESION OF β AMYLOID PRECURSOR PROTEIN AND DOPA-
MINE TRANSPORTER. S. Nakamura, S. Sudoh, H. Yamashita, Y-X. Zhang, T. Kawarai, H.
Kawakami, T. Takahashi, S. Kitayama* and T. Dohi* 3rd Dept. of Internal Med., School of Med. and
*Dept. of Pharmacology, School of Denstry Hiroshima University, Hiroshima, Japan 734

We studied changes in the expression of β amyloid precursor protein (APP) by the administration of
nicotine. When (-)nicotine 5mg/kg was administered subcutaneously to Wistar rat, the expression of
APPmRNA in the cerebral cortex after 4 hours was lower than control rats, although the differnce was not
significant. The administration of (-)nicotine 3mg/kg daily for 10 days resulted a lowered expression of
APPmRNA in the cerebral cortex than control rats, although the differnce was not significant. We also
detected a decrease in APPmRNA in cultured SK-N-SH neuroblastoma cells 4 hours after nicotine was
added. A more suitable condition might produce a significant *in vivo* decrease in APPmRNA.

We also studied the effect of nicotine on dopamine uptake, using PC12 or COS cells transfected with
dopamine transporter gene (pcDNAT1). In PC12 cells, dopamine uptake was inhibited by nicotine, while
nicotine did not influence the dopamine uptake in pcDNAT1 transfected COS cells. PC12 cells treated with
β NGF, expressing abundant nicotinic acetylcholine receptor (nAChR) showed a more prominent inhibition of dopamine uptake by lower concentrations (10^{-7} - 10^{-3}M) of nicotine. The pretreatment with 0.5
mM hexamethonium abolished the inhibitory effect of nicotine on dopamine uptake. Nicotine might presumably inhibit dopamine uptake through nAChR. Nicotine or dopamine would regulate the amyloid gene-
sis through intracellular signal transduction, since APPmRNA expression was suppressed by protein kinase C inhibitor and FOS-JUN complex regulates APP gene via AP-1 region in APP promotor gene of APP.

NICOTINE STIMULATION OF NERVE GROWTH FACTOR RECEPTOR EXPRESSION A.V. Terry Jr., and
M.S.F. Clarke Department of Pharmacology and Toxicology (AVT), and Department of Cellular Biology and
Anatomy (MSFC) Medical College of Georgia, Augusta, GA 30912 and Department of Veterans Affairs Medical
Center (AVT), Augusta, GA 30912

A number of previous studies have suggested that nicotine may have beneficial actions in neurodegenerative
disease models including trophic actions on neurons and extended enhancement of cognition. The purpose of the
experiments described in this study was to determine whether the long lasting and beneficial effects of nicotine
observed previously could be expressed through actions upon nerve growth factor (NGF) receptors. Using a
differentiated PC-12 neuronal cell model, we have detected an increase in expression of cell surface NGF receptor
protein after acute exposure to nicotine. Exposure to nicotine at concentrations of 1.0 and 10 μM for 4 hr produced
statistically significant increases (2.3 %) and (4.7 %) respectively in cell surface NGF receptor protein over that
detected in control cultures (i.e cells not exposed to nicotine). Doses of 0.1 μM and below were not found to
produce significant effects after a 4hr exposure. However, after 24 hr exposure to nicotine, doses as low as 10 nM,
caused significant increases (ranging from 8.8 % to 14 %) in cell surface NGF receptor protein. In addition, we
have also observed a persistent effect upon NGF receptor expression which lasted even after nicotine was removed
from the tissue culture medium. In these experiments, the level of NGF receptor protein increased significantly (9
%) in the presence of 10 nM nicotine after an initial 24 hour exposure as previously observed. Interestingly, 24
hr after the removal of nicotine from the tissue culture medium, cell surface NGF receptor protein continued to
increase dramatically (36 %). By 48 hr after the removal of nicotine from the culture medium, NGF receptor
expression had subsequently returned to control levels. Furthermore the increase in cell surface NGF receptor protein
was blocked in the presence of 10 μM mecamylamine, indicating that this effect is likely nicotinic receptor
mediated. These results are consistent with the hypothesis that the lasting and beneficial actions of nicotine
previously observed *in vivo* may involve an indirect effect upon the level of neuronal cell surface NGF receptor
expression. Our observations offer one possible mechanism for the "functional" neurotrophic effect of nicotine.

Section 8: Effects on transmitter release and ion fluxes

P 58

SODIUM CHANNELS CONTRIBUTE TO NICOTINE-STIMULATED Rb^+ EFFLUX IN 12 BRAIN REGIONS. T.K.Booker, A.E.Bullock, M.J.Marks, A.C.Collins. I.B.G., University of Colorado, Boulder.

While binding assays are useful in determining numbers of nicotinic sites in the brain, they provide little information about the functional status of these putative receptors. In order to provide a general biochemical measure of receptor function, we have recently developed a Rb^+ efflux assay. Synaptosomes prepared from mouse brain tissue are loaded with $^{86}Rb^+$ ions and then stimulated with nicotine. When the nAchR is activated by nicotine, the ion channel opens, allowing Rb^+ ions to flow out of the synaptosome. The amount of Rb^+ released can then be used to quantitate the functional status of the nAchR. However, ion flux may occur through channels other than those associated with nAchRs. In fact, tetrodotoxin (TTX), a Na^+ channel blocker, inhibits some of the nicotine-stimulated Rb^+ efflux. Different brain regions contain varying numbers of nAchRs. Experiments were conducted with or without 100nM TTX to determine if TTX had differential effects on Rb^+ efflux in 12 separate regions. TTX uniformly inhibited Rb^+ efflux by 48% in all regions (r=.994). Rb^+ efflux and nicotine binding strongly correlate in these brain regions (r=.903). TTX treatment reduces Rb^+ efflux by almost half in each region, but the correlation between nicotine binding and Rb^+ efflux remained intact in the presence of TTX (r=.894). Saxotoxin binding was conducted to determine the number of Na^+ channels. Substantially more saxotoxin binding sites than nicotine binding sites were present in the preparation. No correlation between Rb^+ efflux and the number of Na^+ channels was found. These data suggest that the Rb efflux upon nicotine stimulation is due to contributions from several channels, the nAchR and Na^+ channels, and that direct actions of the channel associated with the nAchR may be observed by blocking the activation of the Na^+ channels.

P 59

CONTRIBUTION OF Na^+ AND K^+ CHANNELS TO DESENSITIZATION OF NICOTINE-STIMULATED $^{86}Rb^+$ EFFLUX IN MOUSE STRIATAL SYNAPTOSOMES. A.E. Bullock, M.J. Marks, A.C. Collins, Institute for Behavioral Genetics, Univ. of Colorado, Boulder, CO 80309

Recently we have demonstrated that nicotine-stimulated $^{86}Rb^+$ efflux can be used to monitor nicotinic receptor function in mouse brain synaptosomes. In this study, Na^+ and K^+ channel blockers were used to investigate the contribution of ion channels to nicotine-stimulated $^{86}Rb^+$ efflux and desensitization of this response. Synaptosomes prepared from thalamus of C57BL/6 mice are loaded with $^{86}Rb^+$ and continuously perfused with buffer. Desensitization of nicotine-stimulated efflux was assessed by continuous perfusion with stimulating concentrations of nicotine ($0.1 - 30\mu M$). The peak effect and rate of desensitization of nicotine-induced $^{86}Rb^+$ efflux increased with increasing concentration of nicotine. In the presence of 100nM tetrodotoxin (TTX), a Na^+ channel blocker, a decrease in both nicotine-induced peak $^{86}Rb^+$ efflux and the rate of desensitization of this response was seen indicating that some of the $^{86}Rb^+$ is moving through Na^+ channels. In contrast, stimulation with 10 μM nicotine in the presence of 5 mM CsCl, a potassium channel blocker, had no effect on either maximal nicotine-induced $^{86}Rb^+$ efflux or the rate of desensitization of this response. Both TTX and Cs^+ decreased basal synaptosomal $^{86}Rb^+$ efflux. The time course and maximal recovery from nicotine-induced desensitization was unaffected by TTX and Cs^+. These results demonstrate that Na^+ channels, but not K^+ channels, contribute substantially to both peak effect and the rate of desensitization of nicotine-stimulated $^{86}Rb^+$ efflux in mouse brain synaptosomes. Therefore, desensitization of nicotine-stimulated $^{86}Rb^+$ efflux during blockade of both Na^+ and K^+ channels may provide a more accurate assessment of nicotinic receptor function. (Supported by DA03194, DA00197 and HD07289-09)

P 60

FUNCTIONAL CONSEQUENCES OF DISULFIDE MODIFICATION ON NICOTINE-EVOKED ^{3}H-DOPAMINE RELEASE FROM STRIATAL SYNAPTOSOMES. SR Grady, MJ Marks, and AC Collins, IBG, Univ. of Colorado, Boulder, CO 80309.

Mouse striatal synaptosomes, (after ^{3}H-dopamine (DA) uptake), were treated with 0.1-10 mM dithiothreitol (DTT) or 1 mM dithio(bis)nitrobenzoate (DTNB) for 3 min using a continuous perfusion apparatus. After disulfide or sulfhydryl modification, synaptosomes were exposed to L-nicotine (NIC) or elevated K^+ to evoke DA release. Fractions were collected every 30 s. Data were analyzed by computer-generated curve fits. DTT treatment prevents most NIC- but not K^+-evoked DA release, indicating that DTT modifies function of the nicotinic receptor (nAChR) but not subsequent events in DA release. DTNB by itself has no effect on DA release but reverses the effect of DTT. NIC-induced DA release has two components; the 1st desensitizes rapidly ($t_{1/2} \approx$ 20 s) and has an EC_{50} for NIC of 1 μM, while the 2nd is slow to desensitize ($t_{1/2} \approx$ 10 min) and has a low EC_{50} for NIC (40 nM). At 10 μM NIC the initial ratio of 1st to 2nd component release is 3:1. Both components are completely blocked by 10 μM mecamylamine or dihydroßerythroidine indicating that both are mediated by nAChRs. DTT affects both components. After treatment with 3 mM DTT, the EC_{50} for 2nd component release is shifted from 40 nM to 2 μM with no effect on maximum release, and the 1st component NIC EC_{50} shifts from 1 μM to over 100 μM. For both components, the IC_{50} for DTT varies with the NIC concentration used to evoke release, consistent with the effect on potency rather than efficacy. The K_i calculated for DTT is $\approx$ 0.1 mM for both components. These results indicate that either two forms of nAChR mediate DA release or the two components represent two kinetic aspects of a single receptor. In either case, the 2nd component, which is very slow to desensitize, has interesting implications for the reinforcing properties of nicotine. (Supported by DA03194 and DA00197).

CONTRASTING EFFECTS OF ADRENALECTOMY ON NICOTINE-INDUCED DOPAMINE RELEASE AND LOCOMOTOR DEPRESSION IN RATS.

Mohammed Shoaib* & Toni S. Shippenberg. Drug Abuse Group, Max-Planck Institute of Psychiatry, Clinical Institute, 80804 Munich, Germany.

Adrenalectomy (ADX) in mice can potentiate physiological and behavioural responses to nicotine (Pauly, Ullman and Collins 1990; Pharmac Biochem Behav 35:171-179). The present experiments sought to examine these observations in rats and to extend the measures to dopamine-releasing properties of nicotine. Dose-dependently nicotine (0.4-1.2 mg/kg s.c.) depressed locomotor activity, an effect that was potentiated by ADX. However, in rats chronically treated with nicotine for 5 days (daily injections of 0.4 mg/kg sc), the locomotor depressant effects of nicotine did not differ from saline treated controls. Using in vivo microdialysis, nicotine (0.4 mg/kg s.c.) increased extracellular dopamine levels by 174% in the nucleus accumbens, but this response remained unaffected in rats pretreated with nicotine for 5 days (daily injections of 0.4 mg/kg s.c.). However, ADX rats showed much smaller increases in dopamine following nicotine administration; in both saline and nicotine chronically treated groups, nicotine increased dopamine levels by 140% from baseline conditions. This attenuation by ADX was confirmed when the results were analysed using absolute concentrations of dopamine. The amount of dopamine released was greater in rats treated chronically with nicotine. However, in ADX rats this sensitised response was diminished. The results suggest that the adrenal glands are important in modulating sensitivity to nicotine. Despite ADX potentiating the locomotor depressant effects of nicotine, this manipulation failed to potentiate nicotine-induced dopamine release. Therefore, it suggests that endogenous corticosteroids as well as modulating sensitivity to nicotine may also exert an effect on the mesolimbic dopamine system. *Supported by The Wellcome Trust, United Kingdom.

ISOARECOLONE AND NICOTINE: COMPARING INDUCED DOPAMINE RELEASE IN THE FRONTAL CORTEX AND STRIATUM OF DRUG NAIVE AND NICOTINE TREATED RATS.

P. Whiteaker and S. Wonnacott, School of Biology and Biochemistry, University of Bath, Bath, BA2 7AY, U.K.

Isoarecolone is a nicotinic agonist, known to have some different behavioural actions to nicotine[1] Using a superfusion apparatus similar to that previously employed in this laboratory[2], the abilities of nicotine and isoarecolone to induce dopamine release in synaptosomal preparations from two brain regions were compared. Using drug naive animals, dose response relationships were examined for each of the compounds. In the striatum, nicotine (in the range 0.01μM-150μM) elicited a complex response. High levels of non-specific (mecamylamine insensitive) release were seen at high nicotine concentrations, and the specific nicotine response was biphasic with peaks at 5μM and 90μM. In contrast to nicotine, isoarecolone evoked release in the range 1μM-170μM, producing minimal non-specific release, and apparently monophasic specific release: isoarecolone was both less potent, and less efficacious. Release from the frontal cortex was consistent with mediation by a single nicotinic site, but again isoarecolone induced less non-specific release, and was less potent and efficacious. Release was also measured in animals chronically dosed with nicotine (4mg/kg/day, 14days) using implanted osmotic pumps. This treatment lead to increased [^{3}H]nicotine binding density, but caused a paradoxical drop in response to subsequent nicotinic challenges, in line with recent findings in the mouse[3]. This study suggests a basis for the observed disparities between the behavioural effects of isoarecolone and nicotine.

1). Reavil, C. et al., (1990) Psychopharmacology 102: 521-528.
2). Rowell, P.P. et al., (1990) J. Neurochem. 55: 2105-2110.
3). Marks M.J. et al.. (1994) J. Pharmacol. Exp. Ther. 266: 1268-1276.

We gratefully acknowledge the M.R.C., the Wellcome Trust and Ian Stolerman.

NIFEDIPINE INHIBITION OF NICOTINE-EVOKED [^{3}H]DOPAMINE RELEASE FROM RAT STRIATAL SYNAPTOSOMES IS NOT MEDIATED BY L-TYPE CALCIUM CHANNELS.

Richard J. Prince and Patrick M. Lippiello, R.J.Reynolds Tobacco Co, Pharmacology Div., Winston-Salem NC 27102

Whilst previous studies have suggested that L-type calcium channels are not involved in the nicotine-evoked release of dopamine[1], we have observed that the L-type antagonist, nifedipine, is capable of modulating the release of [^{3}H]dopamine from superfused rat striatal synaptosomes evoked by 10μM nicotine or 40mM K$^+$. Nifedipine completely abolished nicotine-evoked release with an IC$_{50}$ of 22μM. K$^+$ evoked release was inhibited with a similar IC$_{50}$ but the maximal extent of inhibition was around 30%. As nicotine gives 20-30% of the response evoked by 40mM K$^+$, these results suggest that the same portion of the total releasable pool has been inhibited in both cases. In contrast, the L-type channel activator, BAY-K8644 at a concentration of 10μM was without effect alone, or in the presence of nicotine or K$^+$. The concentration of nifedipine required to inhibit striatal dopamine release is 100-1000 fold higher than that required to block L-type channels[2]. This, together with the lack of effect of BAY-K8644 suggests that L-type channels are not involved in evoked release. Recently, it has been shown that dihydropyridines can directly inhibit the nicotinic receptor of bovine chromaffin cells[3]. Whilst direct interactions cannot be ruled out, the inhibition of K$^+$ evoked responses suggests that another mechanism may be operating. We speculate that nifedipine may be acting either upon some other calcium channel subtype, or upon a post-depolarization component of the release process.

1) Harsing et al., Neurochem. Res.17, 729-734 (1992)
2) Adamson et al., Eur. J. Pharmacol. 162, 59-66 (1989)
3) Lopez et al., Eur. J. Pharmacol. 247, 199-207 (1993)

P 64   LOBELINE EVOKES A MARKED UTILIZATION OF DOPAMINE IN RAT STRIATUM

L.P. Dwoskin, L.H. Teng and P.A. Crooks. College of Pharmacy, University of Kentucky, Lexington, KY, 40536 USA.

Lobeline (α-lobeline) is a constituent of Indian tobacco (*Lobelia inflata*), which displaces [³H]nicotine binding at nicotinic receptors in brain. At high doses, lobeline produces discriminative stimulus effects in rats trained in a T-maze; however, no evidence was provided to link this effect to nicotinic receptor activation. The present study examined the effect of lobeline (0.01 - 100 μM, as the sulfate salt) and nicotine (0.05 -100 μM, as the ditartrate salt) on [³H]dopamine ([³H]DA) release from rat striatal slices superfused in the presence of nomifensine (10 μM) and pargyline (10 μM). Lobeline and nicotine increased· [³H]DA release in a concentration-dependent manner. At low concentrations the effect of lobeline was 30 - 50% of the response to nicotine, whereas at high concentrations the effect of lobeline was 10-32 fold greater than that of nicotine. As a result, tissue content levels of DA and dihydroxyphenylacetic acid (DOPAC) was assessed after exposure to high concentrations of lobeline. Lobeline (30 and 100 μM) depleted DA content to 60% and 41% of control, respectively, and furthermore increased DOPAC content 2.9- and 4.2-fold, respectively, compared to control. Taken together, the findings that lobeline increases DA release, depletes DA content and increases DOPAC content in striatal slices suggest that lobeline markedly enhances DA utilization, and potentially induces neurotoxicity in the nigrostriatal dopaminergic system. Since lobeline is an extremely lipophilic drug, it has the potential to accumulate in brain following chronic administration. Current studies are determining the effect of chronic low dose administration and potential toxicity. Thus, these findings may be relevant to the evaluation of lobeline as a smoking cessation product. (Supported by a grant from the Tobacco and Health Research Institute, Lexington, KY.)

P 65   **NICOTINE-INDUCED DOPAMINE OVERFLOW IN THE NUCLEUS ACCUMBENS: EVALUATION OF POSSIBLE MECHANISMS OF ACTION.** Enrico Museo and Agu Pert. Biological Psychiatry Branch, National Institute of Mental Health, Bethesda, MD 20892.

The psychomotor stimulant effects of nicotine are believed to be due, at least in part, to the activation of a dopaminergic substrate; one dopamine (DA)-containing region that appears to be involved is the nucleus accumbens (nAcb). We previously reported that the systemic or intra-nAcb administration of nicotine stereoselectively facilitates DA overflow in this region. Various hypotheses have been proposed to explain this facilitative action on DA overflow. We tested two hypotheses: 1) that nicotine acts as a reuptake blocker, and 2) that nicotine's actions are dependent on the integrity of glutamatergic transmission. Microdialysis probes were positioned in the nAcb of anaesthetized rats and various pharmacological agents were administered through the probe. The test of the first hypothesis was based on the fact that reuptake blockers interfere with the DA-releasing actions of amphetamine: whereas the classic reuptake-blocker nomifensine (1μM) attenuated the effects of amphetamine (0.5 μM), nicotine (100 μM) failed to do so. These findings suggest that nicotine's actions on nAcb DA overflow are not likely due to the blockade of DA reuptake. The second hypothesis was tested using NMDA antagonists; the intra-nAcb administration of either the noncompetitive antagonist MK-801(100 μM) or the competitive antagonist AP-5 (100 μM) prior to and during the administration of nicotine (100 μM) reduced the magnitude of nicotine's effects on DA overflow. It appears that the local actions of nicotine on DA overflow in the nAcb are dependent on the integrity of glutamatergic transmission.

P 66   COMPARISON OF THE EFFECTS OF CONSTANT NICOTINE INFUSION ON NUCLEUS ACCUMBENS AND STRIATAL DOPAMINE RESPONSES TO ACUTE NICOTINE M.E.M. Benwell and D.J.K. Balfour Dept of Pharmacology, University Medical School, Ninewells Hospital, Dundee, Scotland. DD1 9SY.

Nomifensine (5μM), added to the microdialysis perfusion solution, unmasks the stimulatory effects of acute nicotine boli on dopamine (DA) overflow in the nucleus accumbens (NAc) [1]. These conditions have been used to compare the effects of acute nicotine boli and constant nicotine infusions on DA secretion in NAc and striatum (STR). Over the dose range tested, bell-shaped responses were obtained in both brain regions such that 0.1, 0.4 and 0.8 mg/kg sc injections of nicotine resulted in extracellular DA levels of 146±15, 151±9 and 138±10% in the NAc and 98±17, 137±27 and 119±5% in the STR respectively when expressed as a percent of baseline levels measured prior to injecting the nicotine bolus. The response in the NAc was significantly (p<0.01) greater than that in the STR. The constant infusion of nicotine (1mg/kg/day) significantly attenuated (p<0.01) the DA response elicited by an acute bolus of nicotine (0.4mg/kg sc) in the NAc from 145±9 to 105±8% but had no effect on the response in the STR (123±6 and 126±9% in the saline- and nicotine-infused rats respectively). These data support the conclusion that the mesolimbic DA system is more sensitive than the nigrostriatal DA system to both the stimulant effects of a nicotine bolus and the desensitision of the response caused by constant infusion of the drug.

1. Benwell, M.E.M. & Balfour, D.J.K. (1992) Br. J. Pharmacol., 105, 849-856.

This study was supported by Verum Foundation.

DESENSITISATION OF THE STIMULANT EFFECTS OF NICOTINE ON NORADRENALINE SECRETION IN RAT HIPPOCAMPUS M.E.M. Benwell and D.J.K. Balfour Dept of Pharmacology, University Medical School, Ninewells Hospital, Dundee, Scotland. DD1 9SY.

Previous studies in our laboratory have shown that the constant infusion of nicotine, at a dose which maintains plasma levels of the drug at a concentration (24 ± 5 ng/ml) commonly found in habitual smokers, inhibits the stimulatory effects of a nicotine injection on dopamine (DA) overflow in the nucleus accumbens. The study presented here used in vivo microdialysis to investigate the effects of nicotine, constantly infused at the same rate (1mg/kg/day), on the increase in noradrenaline (NA) overflow in the ventral hippocampus (coordinates for dialysis probe were 5.8mm caudally, 4.9mm laterally and 7mm vertically from Bregma) evoked by nicotine injections. The administration of an acute nicotine bolus (0.4 mg/kg sc) increased ($p<0.05$) NA overflow to $164 \pm 19\%$ of the pre-nicotine bolus baseline. This response was enhanced to $258 \pm 52\%$ if the animals were given 5 daily injections of the drug prior to the test day. The constant infusion of nicotine abolished ($p<0.05$) the response to an acute injection of nicotine and attenuated ($P<0.05$) the peak NA overflow to $140 \pm 19\%$ in the sensitised rats pretreated with daily injections of the drug. The data support the conclusion that, in common with the effects of nicotine on the mesolimbic DA system, the receptors which mediate the stimulatory effects of nicotine, on hippocampal NA systems, may be desensitised by exposure to levels of nicotine which are likely to occur in the brains of habitual smokers.

This study was supported by Verum Foundation.

EFFECT OF NICOTINE AND NICOTINIC AGONISTS ON RAT BRAIN BIOGENIC AMINES. K. Summers and E. Giacobini. Dept. Pharmacology, Southern Illinois University School of Medicine, P.O. Box 19230, Springfield, Illinois 62794-9230 USA.

Transcortical microdialysis was employed to investigate the effect of systemic and local (through the probe) administration of nicotine and nicotinic agonists on cortical extracellular levels of acetylcholine (ACh), norepinephrine (NE), dopamine (DA) and serotonin (5-HT). No cholinesterase inhibitor was added to the perfusate. No changes in EEG were observed. Systemic administration of [-]-nicotine produced a 106% increase of cortical ACh release over basal levels that persisted for approximately 2 hr. Concurrently, NE levels were increased 86% over basal values for 60 min. The effects appear to be stereoselective, as systemic injections of [+]-nicotine significantly increased cortical ACh levels only 48% over basal levels for 30 min, and NE levels in the dialysate only 43% over control levels for 60 min. No significant changes of basal DA or 5-HT levels were observed, although DA did appear to increase in response to systemic nicotine. L-nicotine (3.6 μmol/kg, s.c.) significantly increased ACh levels (100%) in the dialysate for 2 hrs and significantly increased NE levels (85%) for 1 hr. D-nicotine (3.6 μmol/kg, s.c.) increased ACh and NE (30%) for 1 hr, indicating stereoselectivity. Anabaseine (3.6 μmol/kg, s.c.) appears to be more potent than nicotine at elevating levels of both ACh (150%) and NE (100%) while fluoronicotine (3.6 μmol/kg, s.c.) appears to be less potent at elevating ACh (70%) and NE (20%). No significant effects of these compounds have been observed on DA or 5-HT levels. (Supported by R.J. Reynolds Tobacco Co.)

Section 9: Electrophysiological effects

P 69

CHARACTERIZATION OF NICOTINIC CURRENTS IN CULTURED SYMPATHETIC NEURONS BY PATCH–CLAMP TECHNIQUES. Chuan–gui Liu, Qing–song Liu and Xiang–ping He. Institute of Pharmacology and Toxicology, P.O.Box 130, Beijing 100850, P.R.China.

Responses of cultured rat superior cervical sympathetic neurons to local applications of cholinergic drugs were studied using whole cell, cell attached and outside–out patch versions of the patch clamp technique. All neurons tested were activated by three nicotinic agonists with the following rank order: DMPP > Nicotine > ACh. With 100 μM each of these agonists, the peak responses ranged from 100 pA to 4 nA at a membrane potential of -60 mV. The nicotinic responses were blocked by hexamethonium, mecamylamine, curare, but not by α–bungarotoxin. The blockade of mecamylamine was use–dependent. Whole cell currents evoked by the three nicotinic agonists showed strong rectification at positive membrane potentials. The nicotinic responses declined progressively in the presence of agonist, a phenomena was known as desensitization. The rate of desensitization was both dose and voltage–dependent. The single nicotinic channels are studied using DMPP as an agonist. In outside– out patch, simultaneous openings of multi– channels was immediately detectable with the onset of DMPP perfusion. In cell– attached patch, DMPP activated single channel events of many different conductance: 15 ps, 27 ps and 38 ps, and with the 27 ps being the most frequently seen conductance. DMPP–activated single channels generally have two gating modes, the brief opening mode and a burst of long opening mode. The open and closed durations are distributed according to double exponentials. After 2 weeks in culture, some neurons showed spontaneous synaptic currents, which could be blocked by hexamethonium. It provides functional evidence for a cholinergic nature of synaptic transmission between cultured sympathetic neurons.

P 70

THE FUNCTIONAL ROLE OF NICOTINIC RECEPTORS IN THE RAT PREFRONTAL CORTEX : ELECTROPHYSIOLOGICAL, BIOCHEMICAL AND BEHAVIORAL CHARACTERIZATIONS. C. Vidal. Laboratoire de Neurobiologie Moléculaire, Institut Pasteur, 25 rue du Dr Roux, 75724 Paris Cedex 15, France.

Animal and human studies have shown that nicotine improves learning and memory in a variety of behavioral tasks. However, the mechanisms and sites of action of nicotine in promoting cognitive functions are still largely unknown. In electrophysiological studies performed in the prefrontal cortex of the rat, we showed that nicotine facilitates glutamatergic neurotransmission : intracellular recordings in pyramidal cells revealed that nicotine selectively enhances the amplitude of excitatory postsynaptic potentials, possibly via presynaptic nicotinic receptors located on glutamatergic terminals. Experiments using *in vivo* microdialysis are consistent with this view. The analysis of dialysates collected in the prefrontal cortex of awake rats showed that the basal concentrations of glutamate were increased by the local application of nicotine in a dose-dependent manner. Thus, the facilitation by nicotinic receptors of glutamatergic synapses in the prefrontal cortex might influence the processing of information in this area. The behavioral correlates of the above findings were investigated by examinating the effects of microinjections of nicotinic antagonists in the rat prefrontal cortex on spatial working memory tasks tested in a T-maze. We observed that the local injection of the nicotinic blocker, neuronal bungarotoxin, decreased performance scores in the matching to sample task but not in the non-matching to sample task. The results suggest that nicotinic receptors in the prefrontal cortex of the rat might be of particular importance in working memory tasks involving attentional processes.

P 71

INTRACELLULAR AND SINGLE CHANNEL STUDIES OF NEURONAL NICOTINIC ACETYLCHOLINE RECEPTORS IN CHICK BRAIN SLICES. W.R. Weaver and V.A.Chiappinelli. Dept. of Pharmacological and Physiological Science, Saint Louis University, St. Louis, MO 63104.

We have examined central neuronal nicotinic acetylcholine receptor (NNAChR) heterogeneity in the lateral spiriform nucleus (SPL) of chick embryos (17-21 days of incubation). These receptors have a high affinity for nicotinic agonists and are insensitive to blockade by kappa- or alpha-bungarotoxin (Neuron 5, 1990; 307-315). Using intracellular recordings in brain slices and perfusions with multiple concentrations of agonist (3 μM - 3 mM carbachol with 1 μM atropine) in the presence of various concentrations of antagonists (0-500 μM), we find that trimethaphan inhibits SPL nicotinic responses with a wide range of K_i's (4-66 μM), while dihydro-β-erythroidine (DHβE) inhibits these responses in a very narrow concentration range (K_i's = 0.09-0.16 μM). These results demonstrate NNAChR heterogeneity in the SPL and suggest that trimethaphan distinguishes between these receptor subtypes. We are now examining single channel responses to acetylcholine in chick brain slices to determine if multiple subtypes of NNAChR channels can be identified and characterized in the SPL. Preliminary results indicate the presence of different individual channels in cell-attached patches that may be distinguished on the basis of slope conductances and channel open times. Single channels which open in response to acetylcholine (1-30 μM) have reversal potentials between -25 and +30 mV, slope conductances from 13 to 89 pS, and open times between 1 and 100's of msec. Supported by NIH Grant NS17574.

THE MECHANISM UNDERLYING THE NICOTINIC HYPERPOLARIZATION OF RAT
DORSOLATERAL SEPTAL (DLSN) NEURONS Eva M. Sorenson and Joel P. Gallagher Dept. of
Pharm., University of Texas Medical Branch, 301 University Blvd., Galveston, TX 77555-1031

Using intracellular recording, we have demonstrated that nicotine and DMPP hyperpolarize DLSN
neurons by a postsynaptic mechanism since the hyperpolarization persists in TTX, low Ca^{2+}, or after
treatment with 1 mM dithiothreitol to inactivate excitatory nicotinic receptors (Sorenson and Gallagher,
1992). The response is blocked by mecamylamine (≥ 50 μM) and kappa-bungarotoxin (0.5 μM) but
not alpha-bungarotoxin (1μM). Opening of a K^+ channel mediates the response since high K^+ buffer
shifts the reversal potential in a Nernstian manner. We have now examined the coupling mechanism
between the nicotinic receptor and the K^+ channel involved. Intracellular injection of BAPTA abolished
the nicotinic hyperpolarization while sparing the $5HT_{1A}$ response in the same neurons. A minimal level
of intracellular Ca^{2+} is therefore needed for the response. The response persisted in 0 mM Ca^{2+}, 1 mM
EGTA buffer, thus influx of extracellular Ca^{2+} is not required. In the absence of extracellular Ca^{2+},
thapsigargin (1-5 μM), which depletes IP_3-sensitive Ca^{2+} stores, did not block the hyperpolarization.
To test the involvement of G proteins, GTPγS was injected into the neurons. In the 3/3 cells in which
GTPγS abolished the $5HT_{1A}$ response, the response to DMPP was also blocked; therefore, G proteins
may be involved. Since ≥ 1 μM of apamin was needed to block the hyperpolarization, the K^+ channel
involved is not an SK channel. In summary, the inhibitory nicotinic response requires a minimal level
of intracellular Ca^{2+}. However, Ca^{2+} influx into the cell or release of Ca^{2+} from IP_3-sensitive stores
is not essential. The nicotinic receptor is coupled to a K^+ channel through a G protein. Although the
K^+ channel appears to be Ca^{2+}-dependent, it is not a classical SK channel. (Supported by DHHS 1F32
DA 05447 to EMS and The Council for Tobacco Research, USA, Inc. to JPG)

EFFECTS OF NICOTINE ON THE FIRING ACTIVITY OF DORSAL RAPHE SEROTONERGIC
(5-HT) NEURONES IN THE RAT. M. Hajós, T. Sharp and G. Engberg* Oxford University
Department of Clinical Pharmacology, Radcliffe Infirmary, Woodstock Road, Oxford, OX2 6HE,
U.K. and * Department of Pharmacology, University of Göteborg, 413 90 Göteborg, Sweden

The dorsal raphe nucleus has recently been found to be rich in high affinity nicotine binding sites,
indicating the possible involvement of 5-HT neurones in various central effects of nicotine. However,
information on the effects of nicotine on the 5-HT neurotransmission is very limited. Here, we report
results of an ongoing *in vivo* electrophysiological investigation to analyse the effect of nicotine on 5-
HT neurones located in the dorsal raphe nucleus of the rat. Male S-D rats were anaesthetised with
chloral hydrate (400 mg/kg) and a lateral tail vein was cannulated for drug administration.
Extracellular recordings were performed using conventional electrophysiological methods. The
position of each cell was marked and verified histologically. Presumed 5-HT neurones displayed a
slow, regular spontaneous firing, a prominent positive then negative waveform and an action potential
duration of 1 to 2 ms. Furthermore, all cells of this type were inhibited by 8-OH-DPAT, a 5-HT
autoreceptor agonist. The majority of the 5-HT neurones tested (33 of 39) were inhibited by
intravenously administered nicotine in low doses (50-200 μg/kg nicotine hydrogen tartrate) although
they displayed great variability in the degree and duration of this response. The lowest dose tested (50
μg/kg nicotine) inhibited 18 of the 39 recorded cells; the average inhibition was 39%. The inhibitory
effect of nicotine was relatively shortlasting and it was antagonised with mecamylamine but not
chlorisondamine (n=6). Furthermore, high doses of nicotine (1-2 mg/kg, s.c.) increased firing activity
of the recorded 5-HT neurones. In contrast, non-5-HT neurones recorded in the same region (n=5)
responded with transient excitation to low doses of nicotine (50-200 μg/kg). We propose, that nicotine
has dual effect on raphe 5-HT neurones: inhibition in low doses and, in higher doses, activation.

ACUTE NICOTINE ADMINISTRATION DIFFERENTIALLY AFFECTS MESOLIMBIC AND
MESOCORTICAL DOPAMINE ACTIVITY. M. Nisell, G.G. Nomikos, P. Hertel and T.H
Svensson. Dept. of Physiology and Pharmacology, Karolinska Institutet, S-171 77
Stockholm, Sweden.

Activation of central dopamine (DA) activity is considered to be of major importance for
the reinforcing properties of nicotine (NIC). In the present study we have examined the
effects of acute, systemic NIC administration on the mesolimbocortical DA system by
employing *in vivo* extracellular single-cell recordings in anesthetized rats and microdialysis
in freely moving rats. NIC (6, 12 and 25 μg/kg, i.v.) dose-dependently increased firing rate
and burst firing of DA-neurons in the ventral tegmental area (VTA). The increase in burst
activity was significant only in DA-neurons located in the paranigral subdivision of the VTA,
from which subcortical, mesolimbic regions receive their dopaminergic innervation, but not
in DA-cells located in the parabrachial pigmented subnucleus which, in contrast, largely
project to cortical regions. NIC (0.5 mg/kg, s.c.) also increased extracellular concentrations
of DA slightly more in the nucleus accumbens (NAC, 55%) than in the medial prefrontal
cortex (MPC, 40%). Also, the NIC induced acute increase in DA levels lasted longer in the
NAC (200 min) than in the MPC (80 min). Consequently, acute, systemic NIC
administration differentially affects subcortical and cortical dopaminergic activity. Our
findings may relate to both the reinforcing and the cognitive enhancing properties of NIC.

P 75
CHRONIC NICOTINE ADMINISTRATION AND UNILATERAL AMPA LESIONS OF NUCLEUS BASALIS MAGNOCELLULARIS (NBM) INCREASE CORTICAL NEURONAL SENSITIVITY TO NICOTINE. F.A. Abdulla, M.-R. Calaminici, S. Wonnacott, J.A. Gray, J.D. Sinden and J.D. Stephenson, Dept. of Pharmacology, University of Alberta, Canada, Depts. of Neuroscience and Psychology, Institute of Psychiatry, London, and Dept. of Biochemistry, University of Bath, Bath, U.K.

Chronic nicotine administration upregulates cortical nicotinic receptors, but the functional consequences of this effect are unknown. Moreover, it has been shown that nbm lesions increase the behavioural sensitivity to nicotine. The present study investigated the effects of chronic nicotine treatment and of unilateral AMPA lesions of nbm on the sensitivity of frontal cortical neurones to nicotine applied iontophoretically. (-)-Nicotine tartrate (2mg/kg in 0.9% NaCl twice daily for 10 days) was administered to a group of rats and the control received the vehicle. A second group received a unilateral nbm lesions by local injection of S-AMPA, naive rats served as control. Twenty four hours after the last nicotine injection, the nicotinic binding sites was increased, from 1.8 to 2.9 pmol/g wet weight, and so was the proportion of cortical neurones responding to nicotine, from 32.3% to 60%. After unilateral lesions, the density of AChE-positive fibres was decreased by more than 90% and the proportion of cortical neurones responding to nicotine increased from 32.3% to 53.8%. The 2 treatments, chronic nicotine and nbm lesions, also shortened the response latency and prolonged its duration. Responses to glutamate were unaffected by either of the procedures. The results show that the increase in $[^3H]$nicotine binding produced by chronic nicotine administration is associated with an increased functional response to nicotinic and suggest that the increased neuronal sensitivity to nicotine after nbm lesions may also be due to a functional upregulation of nicotinic receptors.

Section 10: Cardiovascular, autonomic and immune effects

P 76
CARDIOVASCULAR EFFECTS OF NICOTINE INFUSIONS. S.L. Cruz[1] and J.E. Villarreal[2]. [1]Sección de Terapéutica Experimental, Depto. de Farmacología, CINVESTAV, IPN. [2]Departamento de Farmacología, UNAM. Apartado Postal 22026, México, 14000, México.

Recently, several products have been introduced to human therapy as adjuncts for smoking cessation which deliver nicotine slowly to avoid its cardiovascular effects as a part of the treatment of nicotine abstinence. This work examines the effect of nicotine slow infusions on the cardiovascular system of non-anesthetized spinal rats. Nicotine (200 µg/kg i.v.) was administered in two modes: a) continuous infusions of different duration: 7.5, 15 or 30 min; and b) single fast bolus. Slow nicotine infusions attenuate the cardiovascular effects of this drug. The lack of peak effects was more apparent with the slowest infusions. In spite of this reduced response, a subsequent bolus injection of 200 µg/kg produced marked increases in blood pressure and moderate tachycardic responses. Therefore, slow nicotine infusions do not prevent the acute sympathetic cardiovascular stimulant effects of subsequent rapid administrations. Overall, these results indicate that nicotine bolus in cigarette smoking can produce important cardiovascular effects in subjects wearing devices that deliver slow nicotine administrations. The issue of the cardiovascular effects of cigarette smoking in patients with transdermal nicotine patches must be carefully studied.

COTININE MODULATES THE CARDIOVASCULAR EFFECTS OF NICOTINE. R. Chahine. P 77

Faculty of Medical Sciences II, Lebanese University, BP 55564, Beirut, Lebanon.

It is well known that chronic smokers respond acutely with blood pressure elevation during smoking, but their blood pressure drops markedly below baseline levels of comparable non smokers during short-time abstinence. Likewise, during chronic nicotine application there is a reversal of the initial hypertension which cannot be explained by tolerance development alone. Thus, concurrently with the acute sympathoexcitatory and pressor effect of nicotine an unidentified depressor mechanism should be activated. It is proposed that cotinine, the major metabolite of nicotine which is increased after nicotine administration and has a longer half life, may be implicated in this paradox. This hypothesis was evaluated on arterial blood pressure and isolated aorta of rabbits. At a high dose of cotinine (100 mg/kg i.v.), a rapid and lasting hypotension was observed. Besides, cotinine pretreatment completely blocks the hypertensive action of high dose nicotine (0.125 mg/kg i.v.). Furthermore, cotinine relaxes the isolated rabbit aorta and induces a dose dependent inhibition of the contractile effects of nicotine, with respect to their physiological ratio. Using pA_x calculation, the slope of regression lines of (x-1, where x is the [nicotine] / -log [cotinine]) is different from 1 (p < 0.05) suggesting an insurmountable antagonism probably via an irreversible inactivation of the receptors by cotinine. The high doses of nicotine and cotinine used in this study are beyond those arising as a result of smoking, however they were selected in order to have a visible response. These observations are in accordance with our previous data that cotinine and nicotine have opposing effects on prostacyclin synthesis and noradrenaline release. It is concluded that cotinine may play a role in controlling and limiting the response to nicotine probably via a vasodilator mechanism.

EFFECTS OF NICOTINE ON CONDITIONED PRESSOR RESPONSES IN NORMOTENSIVE AND HYPER-TENSIVE RATS. T. Kubo, R. Fukumori, K. Taguchi and Y. Hagiwara. Department of Pharmacology, Machida, Tokyo 194, Japan. P 78

Smoking or nicotine is suggested to have an anti-stress action. In the present study, we recorded blood pressure in conscious, freely moving rats and examined effects of nicotine on stimulus-induced pressor responses. On the day before or two days before the experiments, a cathether was inserted into the femoral artery to directly measure blood pressure. Wistar rats, spontaneously hypertensive rats (SHR) and Wistar-Kyoto rats (WKY) were subjected to classical fear conditioning (30 trials) involving the pairing of a tone conditioned emotional stimulus with an electric footshock unconditioned stimulus (3 mA, 2 sec). Conditioned emotional stimulus caused an increase in blood pressure. The pressor response was inhibited by subcutaneous injection of hexamethonium (10 mg/kg). Diazepam (0.1 and 0.5 mg/kg, s.c.) also inhibited the stimulus-induced pressor response in a dose-dependent manner. Nicotine (0.1 mg/kg, s.c.) slightly but significantly inhibited the pressor response. The conditioned pressor response was exaggerated in SHR compared with WKY. The inhibitory effect of nicotine on the stimulus-induced pressor response was greater in SHR than that of WKY. These results indicate that nicotine diminishes pressor responses induced by emotional stress and this inhibitory effect of nicotine is enhanced in SHR. The conditioned pressor response is mediated by sympathetic vasomotor excitation and activation of the adrenal medulla.

ACTIONS OF SUBCUTANEOUS AND INTRATHECAL NICOTINE ON THE SYMPATHOADRENAL AND SYMPATHETIC EFFERENT SYSTEMS F.J.-P. Miao, P.G. Green, W. Jänig, N.L. Benowitz, J.D. Levine, Depts Medicine and Oral Surgery, UCSF San Francisco, CA 94143-0452A, USA; Physiologisches Institute, Christian-Albrechts-Universität zu Kiel, FRG. P 79

In recent experiments we have demonstrated that subcutaneous (SC) or intrathecal (IT) nicotine inhibits bradykinin-induced plasma extravasation (BK-induced PE) in the knee joint of the rat. Since activation of presynaptic ß2-adrenoceptors of sympathetic postganglionic nerve terminals inhibits BK-induced PE, adrenal medullary epinephrine, which can be released by nicotine, has been believed to be a mediator for the inhibitory action of nicotine on BK-induced PE. However, we found that adrenal medullectomy or blockade of the peripheral nicotinic receptors attenuates only the action of SC nicotine (J. Pharmacol. Exp. Ther. 262: 889-895, 1992) and does not decrease IT nicotine action (J. Pharmacol. Exp. Ther. 264: 839-844, 1993). Based on these observations, we hypothesized that: 1) SC nicotine stimulates the adrenal medulla to release mediators, such as epinephrine which acts on sympathetic postganglionic nerve terminals to inhibit BK-induced PE; 2) IT nicotine produces the inhibition of BK-induced PE through a mechanism which does not directly involve the sympathoadrenal or sympathetic efferent system. In these experiments we tested these hypotheses using pharmacological, biochemical, and electrophysiological approaches. We found that: 1) the dose-response curve for the inhibitory action of SC nicotine on BK-induced PE was shifted to the right by intra-articular perfusion of ICI-118,551 (a ß2-adrenoceptor antagonist) while the dose-response curve for IT nicotine was not significantly affected; 2) SC nicotine (from 10^{-1} mg/kg) increased plasma levels of epinephrine and norepinephrine, while IT nicotine was only effective at doses above 10 mg/kg; 3) SC, but not IT, nicotine increased activities of the sympathetic chain. These results suggest that SC nicotine excites the sympathoadrenal and sympathetic efferent systems thereby inhibiting BK-induced PE in the knee joint. The mechanism(s) underlying the inhibitory action of IT nicotine is not clear but appears not to involve the sympathoadrenal or sympathetic efferent system.[Supported by a California Tobacco-Related Diseases grant]

P 80

NICOTINE-INDUCED BRONCHOCONSTRICTION: ROLE OF AXON REFLEX. L.-Y. Lee and J.-L. Hong. Dept. of Physiology, Univ. of Kentucky, Lexington, Kentucky, 40536-0084, USA.

We have recently reported that aerosolized nicotine evokes airway irritation and coughs in nonsmokers (J. Appl. Physiol. 75:1955, 1993) and stimulation of vagal C-fiber sensory endings in the lungs is likely involved. Indeed, these endings can be activated by nicotine delivered either by inhalation (cigarette smoke) or by injection into circulation in dogs (Lee et al., J. Appl. Physiol. 66:2032, 1989). Since stimulation of the C-fiber afferents is known to elicit bronchoconstriction via both the cholinergic reflex pathway and the local "axon reflex", the present study was carried out to determine the role of these afferents in the nicotine-induced bronchoconstriction in anesthetized adult guinea pigs. Cigarette smoke (10 ml) delivered directly into the lungs over 3 consecutive respirator cycles increased R_L from a baseline of 0.32 ± 0.08 cmH$_2$O/ml/sec to 0.97 ± 0.16 cmH$_2$O/ml/sec (n=5) and decreased C_{dyn} from 0.48 ± 0.03 ml/cmH$_2$O to 0.26 ± 0.08 ml/cmH$_2$O; these responses reached peaks in 10-18 sec after the smoke delivery, and returned toward their base lines after 1-4 min. The immediate bronchoconstriction was eliminated either by removing the smoke particulates (including >99% of nicotine) (n=5, p<0.01) or by a pretreatment with hexamethonium (1-5 mg/kg, iv) (n=6, p<0.01), suggesting that nicotine is the causative agent. The smoke-induced bronchial constriction was not significantly affected by bilateral cervical vagotomy (n=6, p>0.05), but was completely abolished in animals receiving a systemic capsaicin pretreatment (50 mg/kg, sc) which desensitized the C-fiber afferents and abolished the axon reflex (n=5, p<0.01). These results suggested that the axon reflex elicited by activation of bronchopulmonary C-fiber afferents plays a major role in the nicotine-induced bronchoconstriction in guinea pigs. (Supported by NIH grants HL-40369, HL-52172, and UKTHRI grant #5-41066)

P 81

NICOTINIC ACTIVATION OF MULTIPLE TRANSMITTER SYSTEMS FROM INTRINSIC NEURONS OF THE RAT GASTRIC FUNDUS. A. McLaren, C.G. Li and M.J. Rand. Pharmacology Research Laboratory, Department of Medical Laboratory Science, Royal Melbourne Institute of Technology, Australia.

In studies on the action of nicotine on specific transmitter systems, little attention is paid to the possibility of effects on other transmitter systems which may produce confounding results. Even when specific steps are taken to eliminate the consequences of activation of other systems (eg, the use of atropine to block muscarinic receptors in studies on noradrenergic transmission), the possibility of still further, often unsuspected effects remains. The stomach has extrinsic sympathetic, parasympathetic and sensory innervations and an intrinsic enteric nerve plexus in which several peptide and non-peptide transmitters have been identified. In the present study, we set out to determine the mediators of neuronal actions of nicotine on isolated preparations of rat gastric fundus strips under various conditions. When tone was absent or low, nicotine (100 μM) produced a biphasic contraction: the first rapidly developing phase was blocked by atropine and can be attributed to activation of cholinergic nerves; the second phase was absent after treatment with capsaicin and can be attributed to activation of sensory nerves and release of one or more neuropeptides. When tone was present or induced, nicotine produced a further contraction followed by a biphasic relaxation: the first phase of which was due to activation of nitrergic nerves and release of the NO-like transmitter; the second phase was due to release of a VIP-like peptide, which may be a cotransmitter in nitrergic nerves, and in addition a third, so far unidentified mediator contributes to the relaxations. Despite the presence of noradrenergic nerves in the tissue and their well known property of being excited by nicotine, no evidence was found for a noradrenergic component in responses to nicotine.

P 82

NICOTINE-INDUCED RELAXATIONS OF RAT GASTRIC FUNDUS INVOLVES PARATHYROID HORMONE-RELATED PEPTIDE (PTHrP). A. McLaren, C.G. Li and M.J. Rand. Pharmacology Research Laboratory, Department of Medical Laboratory Science, Royal Melbourne Institute of Technology, Australia.

Nicotine (100 μM) induces non-adrenergic, non-cholinergic relaxations of rat gastric fundus strips that are mediated by nitric oxide (NO)-like and vasoactive intestinal polypeptide (VIP)-like transmitters (McLaren et al., 1993). After blocking NO synthesis with L-NAME (100 μM) and inactivating VIP with α-chymotrypsin (1 U/ml), nicotine still induced a small relaxation which was insensitive to suramin, desensitisation to α,β-methylene ATP, or capsaicin, and is thus unlikely to be due to a purinergic transmitter or a sensory neuropeptide. PTHrP acts directly on specific receptors of smooth muscle cells of rat gastric fundus strips to induce relaxation (Mok et al., 1989). The relaxation induced by 30 nM PTHrP, was reduced to about 30% of the control value by the selective antagonist [Asn[10], Leu[11]] PTHrP (7-34)-amide (ALPA, 1 μM). In the presence of ALPA, nicotine-induced relaxations were reduced to about 75% of the control value. In some cases, the concurrent presence of L-NAME, α-chymotrypsin and ALPA abolished nicotine-induced relaxations, but the reduction in nicotine-induced relaxations in the presence of all three antagonists was not statistically different from that in the presence of L-NAME and α-chymotrypsin without ALPA. Furthermore, PTHrP-induced relaxations were abolished by α-chymotrypsin. We conclude that PTHrP is involved in nicotine-induced relaxations of rat gastric fundus, but is probably not a neurotransmitter.

McLaren, A., Li, C.G. & Rand, M.J. (1993) Clin. Exp. Physiol. Pharmacol., 20, 451-457.
Mok, L.L.S. et al. (1989) J Bone Mineral Res., 4, 433-439.

THE ACTIONS OF DMPP ON NORADRENERGIC MECHANISMS DIFFER FROM THOSE OF
NICOTINE IN THE RAT ANOCOCCYGEUS MUSCLE. C.G. Li and M.J. Rand. Pharmacology
Research Laboratory, Department of Medical Laboratory Science, Royal Melbourne Institute of
Technology, Australia.

DMPP (1,1-dimethyl-4-phenyl-piperazimium) is often used as a nicotinic agonist despite the fact that it
has other actions that are not shared by nicotine (NIC): thus, it is as potent as cocaine in blocking the
neuronal uptake of noradrenaline (Allen $et\ al.$, 1972). In rat isolated anococcygeus muscles, DMPP (10 μ
M) regularly produced a slowly developing and sustained contraction resembling that produced by
guanethidine (G, 10 - 30 μM), whereas NIC (10 μM) produced variable and transient contractions.
Contractions to NIC, DMPP, G and field stimulation (FS) were abolished by prazosin (0.1 μM) or by
pretreatment of rats with reserpine, indicating they were mediated by noradrenaline acting on α
$_1$-adrenoceptors. Hexamethonium (HM, 100 μM) or tetrodotoxin (TTX, 1 μM) blocked contractile
responses to NIC, indicating that were due to activation of neuronal nicotinic receptors, but did not affect
contractile responses to DMPP or G, and TTX but not HM blocked responses to FS. During the
contractile response to DMPP, responses to FS were converted from contractions to nitrergically
mediated relaxations: this effect is also produced by G. After the tone was raised by G, NIC and DMPP
produced relaxations that were due to activation of nitrergic nerves and release of the NO-like
transmitter, and these relaxations were blocked by TTX and HM indicating they were due to activation of
neuronal nicotinic receptors. Thus, the predominant action of DMPP on noradrenergic nerves was
guanethidine-like, although its action on nitrergic nerves was NIC-like.
Allen, G.S., Rand, M.J. & Story, D.F. (1972) Br. J. Pharmacol., 45, 480-489.

NICOTINIC ACTIVATION OF NITRERGIC NERVES IN THE ANOCOCCYGEUS MUSCLE: A
MODEL TISSUE FOR STUDYING NITRERGIC TRANSMISSION. M.J. Rand and C.G. Li.
Pharmacology Research Laboratory, Department of Medical Laboratory Science, Royal Melbourne
Institute of Technology, Australia.

The paired anococcygeus muscles are attached to the coccyx and run to the lateral part of the terminal
bar of the anal sphincter but are not part of the enteric system and may be identical to the human $m.$
$levator\ ani$. In some species, the coccygeal attachment is continuous with the retractor penis muscles.
The autonomic innervation of the anococcygeus muscles is by postganglionic pelvic parasympathetic and
lumbar sympathetic nerves. Axon profiles in the muscle have been identified as about 60%
noradrenergic (with neuropeptide Y as a cotransmitter), about 5% cholinergic, and the remaining non-
adrenergic non-cholinergic terminals are probably almost entirely nitrergic. The first demonstrations of
nitrergic transmission in 1989 were in isolated anococcygeus muscle from rats and mice. The isolated
muscle is fully relaxed and field stimulation of the intrinsic nerves elicits contractions mediated by
noradrenaline acting of α$_1$-adrenoceptors, but modulated by a counteracting relaxation mediated by the
nitrergic transmitter, which appears to be a NO-donating adduct. Nicotine mimics these effects, but its
actions are complicated by development of desensitization. Blockade of the nitrergic component results
in enhancement of contractile responses to field stimulation and nicotine. When the noradrenergic
component is blocked and the tone of the muscle is raised, field stimulation and nicotine produce
relaxations that are mediated by the nitrergic transmitter. The development of desensitization to nicotinic
activation of nitrergic nerves is less marked than that of noradrenergic nerves.

A NEW MAJOR TARGET FOR NICOTINE: NITRERGIC NERVES. M.J. Rand, C.G. Li and A.
McLaren. Pharmacology Research Laboratory, Department of Medical Laboratory Science, Royal
Melbourne Institute of Technology, Australia.

Nicotine is well known to activate several different types of neurons in the peripheral and central
nervous systems by acting on nicotinic receptors, not only on neuron cell bodies (eg. autonomic ganglion
cells), but also on nerve terminals. In the past 4 years, it has been shown that nerves containing nitric
oxide (NO) synthase release a NO-like (probably a NO-donating adduct) transmitter (nitrergic nerves) are
widespread in the autonomic, enteric and central nervous systems, where they serve a wide range of
physiological functions. These nitrergic nerves are particularly sensitive to activation by nicotine. At
neuroeffector junctions in the autonomic and enteric systems, activation of nitrergic nerves results in
relaxation of smooth muscle. Thus, autonomic nitrergic nerves subserve penile erection, relaxation of the
urethra during micturition and dilatation of the cerebral vasculature. Enteric nitrergic nerves subserve
relaxation of sphincters (lower oesophageal, pyloric, Oddi's, ileocolonic and internal anal) and the
descending inhibition of circular smooth muscle that is an essential component of peristaltic propulsion
of the contents of the small and large intestine. In the central nervous system, nitrergic nerves are
implicated in long-term depression in the cerebellum which is involved in motor learning and long-term
facilitation in the hippocampus which is involved in memory. The relationship of the effects of nicotine
in activating nitrergic systems to other peripheral and central effects of nicotine remain to be elucidated
comprehensively, but studies so far indicate that the effects of nicotinic activation of cholinergic and
noradrenergic nerves in some tissues are considerably modulated by concurrent nitrergic activation.

RESPIRATORY AND BLOOD GAS RESPONSES TO CIGARETTE SMOKE EXPOSURE IN CONSCIOUS RATS EXHIBIT NICOTINE DOSE DEPENDENCY. R.T. Dowell [1,2], A.A. Houdi [1,3], and J.N. Diana [1,2]. [1]Tobacco & Health Res. Inst., [2]Dept. Physiology, and [3]College of Pharmacy. University of Kentucky, Lexington, KY 40546-0236.

Previous studies on the impact of cigarette smoke on respiratory and blood gas responses utilized anesthetized subjects. Because anesthesia would markedly alter and/or attenuate neural respiratory responsiveness, the present studies were conducted in conscious rats. Animals were treated by nose-only exposure to either air puffs, smoke from a low nicotine cigarette, or smoke from a high nicotine cigarette. After a three-day "break in" period, respiratory and arterial blood gas measurements were conducted on day four. Respiratory rate was measured with a balloon placed within the mesh restraining apparatus. Arterial blood was obtained from a carotid artery cannula. Under control conditions, while restrained, arterial blood gas composition was nearly identical in all groups and unaltered versus home cage conditions (pH = 7.48 ± .005; pO_2 = 103 mmHg ± 3; pCO_2 = 31 mmHg ± 1; n = 13-16). Although not statistically significant, respiratory rate responses to restraint were detectably lower in low nicotine exposed animals (168 b/min ± 15) and still lower in high nicotine exposed animals (129 ± 11) compared with sham exposed animals (186 ± 18). Upon each repeated puff, a marked and consistent decrease in respiratory rate (-55 b/min) was observed in low nicotine exposed animals. High nicotine exposure elicited enhanced bradypnea (-101 b/min). Arterial blood gas composition in response to cigarette smoke was consistent with the respiratory rate responses. For high nicotine exposure, at the end of eight puffs, blood gas values (n = 7) were: pH = 7.16 ± .06; pO_2 = 72 ± 8; pCO_2 = 52 ± 3. Bradypnea, acidosis, hypoxia, and hypercapnia elicited by cigarette smoke exposure in conscious rats are nicotine dose dependent and unaffected by pretreatment with atropine, arginine vasopressin antagonist, and topical anesthesia of the nose/upper airway with lidocaine. (Supported by HL33767, KTRB, and AHA, Ky. Affiliate).

THE EFFECTS OF CHRONIC NICOTINE TREATMENT ON STRESS- OR ETHANOL-INDUCED GASTRIC LESIONS. D. Wong and C.W. Ogle. Department of Pharmacology, Faculty of Medicine, The University of Hong Kong, Hong Kong.

Previous studies have shown that chronic nicotine intake in drinking water dose-dependently worsens cold (4°C)-restraint-induced (stress), as well as 40% ethanol-induced, glandular mucosal damage in rat stomachs. In an attempt to elucidate the process underlying these actions of nicotine and to examine the possibility that a local action due to oral administration of the alkaloid is involved, mini-osmotic pumps containing either 0.9% NaCl (delivering 12 μl/day) or nicotine (delivering 0.224, 1.03 or 1.88 mg/kg/day) were implanted subcutaneously in rats for a period of 10 days. Chronic exposure to nicotine dose-dependently intensified stress-evoked ulcer formation and slightly increased mast cell degranulation. Oral administration of 40% ethanol to animals with implanted nicotine-delivering pumps also worsened the adverse effects of ethanol on mucosal lesion formation when compared to the control animals with implanted saline-delivering pumps; mast cell degranulation by ethanol was also slightly, but significantly, worsened in the alkaloid-treated group. These findings indicate that systemically-delivered nicotine can also increase the severity of both stress- and ethanol-induced mucosal lesion formation in rats. They also confirm the idea that the stress ulcer-intensifying mechanism of the alkaloid is mainly through a systemic action where it also produces chronic blockade in the vagal ganglia, to lead finally to gastric muscarinic receptor supersensitivity; the latter effect is not unlike that seen after denervation. In the case of ethanol-evoked mucosal damage, direct stimulation of the stomach wall ganglia by alcohol possibly results in somewhat similar exaggerated post-vagal ulcerogenic responses as in stress.

THE EFFECTS OF CHOLINERGIC AND ß-ADRENERGIC ANTAGONISTS ON NICOTINE-INDUCED SUPPRESSION OF BLOOD AND SPLEEN LEUKOCYTES IN RATS. S. Knopf, A. R. Caggiula, C. G. McAllister, L. H. Epstein, S. M. Antelman, K. A. Perkins, S. Saylor, and R. Stiller. Departments of Psychology, Psychiatry and Anesthesiology, University of Pittsburgh, Pittsburgh, PA 15260 and Pittsburgh Cancer Institute.

Acute nicotine (NIC) administration (1.32 mg/kg free base, s.c.) to male Sprague-Dawley rats reduces the proliferative response of peripheral blood leukocytes (PBL) to the mitogens concanavalin A (Con A) and phytohemagglutinin (PHA), and the response of splenic leukocytes (SL) to PHA (McAllister, 1994). An initial attempt to define the mechanisms by which NIC exerts its immunologic effects involved the cholinergic antagonists mecamylamine (MEC; 1mg/kg) and hexamethonium (HEX; 5mg/kg). HEX, which is restricted to the periphery, had no effect on the NIC-induced suppression of proliferation. MEC, which has access to both peripheral and central receptors, significantly attenuated NIC's effects on both PBL and SL. In a separate study, the peripheral cholinergic antagonist chlorisondamine (0.1 mg/kg, sc) also did not attenuate NIC's effect on PBL proliferation. In a third study, the ß-adrenergic antagonist propranolol (2 mg/kg) antagonized NIC's suppression of SL proliferation but did not attenuate NIC's effect on PBL. These results suggest that nicotinic-cholinergic receptors are involved in mediating NIC's effect on both PBL and SL, while ß-adrenergic receptors may be important only for NIC's immunologic effect on SL. Supported by DA07546

CLASSICALLY CONDITIONED INCREASE OF WHITE BLOOD CELL COUNTS AND GLUCOCORTICOID SECRETION IN RATS. Buske-Kirschbaum, A., Grota, L., Kirschbaum, C., Bienen, T., Moynihan, J., Ader, R., Hellhammer, D.H. & Felten, D.L., Dept. of Clinical Psychology, University of Trier, FRG and Depts. of Neurobiology & Anatomy and Psychiatry, University of Rochester, USA

It is well documented that immune function can be modulated by behavioral processes, e.g. classical conditioning techniques. Although the underlying mechanisms of conditioned immunomodulation are still unknown, the endocrine system is assumed to play an important role in neuro-immune interactions. The present study was designed to investigate classically conditioned modulation of the immune system and the endocrine system using one learning protocol.

Male Lewis/N rats with previously implanted indwelling femoral artery catheters were randomly distributed in four different treatment groups. Conditioned animals (n=17) were provided with a peppermint odor (CS) which was followed by an infusion of 0.1 mg/kg nicotine bitartrate (US). Control groups were treated similarly, however these animals were provided with peppermint combined with saline infusion (saline control; n=7) or with nicotine, but on a non-contingent basis (unpaired control; n=7). This schedule was followed for four consecutive days. On the test day 5 the previously conditioned group was divided into two subgroups which were either reexposed to the peppermint odor combined with saline infusion (conditioned group; n=10) or remained without reexposure of the conditioned stimulus (CS_0-group; n=7). The saline group and the unpaired control group were treated with peppermint and saline on this day. Plasma glucocorticoids and the absolute number of peripheral mononuclear cells (pMNC) were determined 10 minutes before and 10 and 60 minutes after stimulus presentation on day 3 (acquisition) and day 5 (test day).

Conditioned animals reexposed to the peppermint odor showed significant elevation of glucocorticoid concentration and number of pMNC. Despite the same treatment no alteration in both parameters could be determined in the control groups, e.g. the saline group, the unpaired group or the CS_0-group.

The present data extend previous results of conditioned immunomodulation indicating that modulation of immune function in a Pavlovian conditioning protocol may be accompanied (or influenced) by other conditioned reactions, e.g. endocrine responses.

Section 11: Cognitive effects in animals and man

AN ANIMAL MODEL TO STUDY NICOTINE'S EFFECTS ON COGNITION. N. E. Grunberg, J. B. Acri, and E. J. Popke. Dept. of Medical and Clinical Psychology, Uniformed Services University of the Health Sciences, Bethesda, MD, USA, 20814-4799.

The acoustic startle response (ASR) with pre-pulse inhibition (PPI) is a valuable paradigm to study mechanisms underlying effects of sympathomimetic drugs (Davis et al., 1975; Kokkinidis & Anisman, 1978). PPI also has been proposed as a behavioral model to study information processing and sensory gating in conditions involving thought disturbances (Swerdlow, 1986). We have adopted this paradigm to study effects of nicotine on sensory gating and possibly on attentional processes in rats. We believe that this model provides a valuable complement to human and clinical studies investigating effects of nicotine in dementia patients. Results from a series of our recent experiments indicate that: (1) acute and chronic nicotine have an inverted U-shaped dose-effect on ASR and PPI; (2) effects of nicotine on ASR and PPI are altered in the presence of stress; and (3) effects of nicotine on ASR and PPI appear more pronounced in females than in males. This paper will summarize and integrate the findings from this line of research and will emphasize the value of this paradigm to study nicotine's cognitive effects. In addition, specific uses of this model along with neuroscience and molecular-biological techniques will be discussed.

P 91   DISSOCIATION BETWEEN CHRONIC NICOTINE-INDUCED COGNITIVE FACILITATION AND CORTICAL AND HIPPOCAMPAL NICOTINE BINDING. F.A. Abdulla, E.J. Bradbury, M.-R. Calaminici, S. Wonnacott, J.D. Stephenson, J.D. Sinden and J.A. Gray. Dept. of Pharmacology, University of Alberta, Canada, Depts. of Neuroscience and Psychology, Institute of Psychiatry, London, and Dept. of Biochemistry, University of Bath, Bath, U.K.

The present study was conducted to investigate if chronic nicotine-induced cognitive facilitation is due to changes in nicotinic receptor density. (-)-Nicotine tartrate (2mg/kg), and mecamylamine (1.0 mg/kg) were administered chronically to different groups of rats twice daily for 10 days. A third group received the same dose of nicotine for one day and the vehicle (saline) for 9 days. Beginning 24 h after the final drug injection, the groups were compared to a vehicle control group on acquisition of a hidden platform position in the Morris water maze over 20 trials with 30-min inter-trial interval. The rats were killed 48 h after the last drug injection and their frontal cortex and dorsal hippocampus were rapidly dissected to be assayed for nicotinic binding sites. Chronic treatment with nicotine significantly increased the number of cortical and hippocampal binding sites and improved the rate of learning. Chronic treatment with mecamylamine also significantly increased the number of cortical but not hippocampal nicotinic binding sites and significantly decreased the rate of learning. Nicotine given for one day did not alter cortical or hippocampal binding sites, however, the nicotine group showed a faster rate of learning than the vehicle-treated group but was significantly slower in learning than the rats receiving chronic nicotine treatment for 10 days. There were no correlations between the number of cortical or hippocampal nicotinic binding sites and the rate of learning. The results indicate that chronic nicotine-induced learning facilitation is probably not due to cortical or hippocampal nicotinic receptors up-regulation. Supported by R.J. Reynolds Tobacco Co.

P 92   **THE EFFECT OF NICOTINE ON THE ACQUISITION AND RETENTION OF CONDITIONAL DISCRIMINATIONS.** Mirza N.R. and Stolerman I.P. Department of Psychiatry, Institute of Psychiatry, De Crespigny Park, Denmark Hill, London SE5 8AF

The effect of nicotine on the acquisition and retention of two operant conditional discrimination tasks has been assessed. In both studies Lister-hooded rats (300-350 g) were trained on a FR10 schedule for food reinforcement during daily 15 min sessions. In the first study rats were trained to discriminate flashing lights from darkness. In preliminary experiments rats took 60 sessions to reach 70% accuracy. Two groups (n=8) of rats were chronically injected (s.c) with saline or nicotine (0.4 mg/kg) for 10 days before exposure to the experimental apparatus. The nicotine group was then injected with 0.15 mg/kg throughout shaping and training sessions. A further 4 groups (n=8) of rats were injected (s.c) with saline or nicotine (0.05, 0.15 or 0.4 mg/kg) 10 min before training sessions only. Retention was assessed during drug-free sessions after acquisition. Prior chronic nicotine exposure had no effect on acquisition. However, nicotine (0.15 mg/kg) injected before training sessions enhanced choice accuracy over sessions (P<0.002). During retention nicotine (0.05 and 0.15 mg/kg) injected rats were significantly (P<0.05) more accurate. In a second study rats were trained to discriminate combinations of visual (flashing lights or darkness) and tactile (wood or grid floor) cues. In preliminary experiments rats took 40 sessions to reach 90% accuracy. Groups (n=10) of rats were injected (s.c) with saline or nicotine (0.05, 0.15 and 0.4 mg/kg) 10 min before training sessions. As in the first study nicotine (0.15 mg/kg) improved accuracy over sessions (P<0.02). Unlike the first study there was no effect on retention. Thus nicotine consistently improves acquisition of conditional discriminations. Facilitation with nicotine has not previously been shown in operant tasks assessing memory. These results add to a growing body of evidence that nicotine can improve acquisition of cognitive tasks. (N.R. Mirza is a MRC-SmithKline Beecham collaborative student).

P 93   THE COMPARISONS AVOIDANCE AND T-MAZE LEARNING OF RATS AFTER NICOTINE AND LITHIUM ADMINISTRATION. K. Nagai and H. Iso. Dept. of Pharmacology, and Psychology, Hyogo College of Medicine, Hyogo, Japan 663.

We studied the effects of nicotine and lithium on two different types of learning in rats; shuttle avoidance and T-maze.

First, male rats of WKY (Wistar Kyoto rats) strain were given nicotine (N) and lithium chloride (LiCl) or both of them mixed with drinking water (0,01% and 0,075%, respectively, adjusted pH of 7.4) for 2-3 weeks. Then, animals were tested by a 50-trial of shock- avoidance session using a flicker (5Hz) light as CS and 1mA scrambled shock as US. Avoidance performance was different among drug treatments and control (not treated); Lithium (Li) slightly enhanced, but N suppressed avoidance learning. Moreover, both Li and N treatments increased evidently the latency and decreased number of avoidances significantly, so that it showed suppression of learning more than N single treatment. On the other hand, the results of the T-maze learning, one part of which has shown in the earlier report, were contrary to the avoidance data. Thus, Li treatments inhibited remarkably, which was shown by increasing of latency and errors, but N decreased the latency and errors. It means that N enhances, but Li reduces the T-maze learning. And also, both Li and N treatments affected to the learning behavior obstructively or competitively in each other.

These contradictory phenomena of the learning of rats might be explained by the motivational changes induced by drug treatment, N enhanced both hunger and fear, but Li reduced both of them. As the biological bases of such behavioral changes , it is possible to estimate that the effects of N and Li might interfere each other, on the basis of biochemical process of drug action, such as drug receptor information systems in the brain.

STUDIES OF (-)-NICOTINE ON CARBON MONOXIDE-INDUCED LEARNING IMPAIRMENT IN
MICE. [1]M. Hiramatsu, [1]H. Satoh, [1]M. Murai, [1]T. Kameyama and [2]T. Nabeshima, [1]Dept. of Chem.
Pharmacol., Fac. of Pharmac. Sci., Meijo Univ., Nagoya 468 and [2]Dept. of Neuropsychopharmacol. and
Hospital Pharm., Nagoya Univ. Sch. of Med., Nagoya 466, Japan.

The effects of nicotine on carbon monoxide (CO)-induced learning impairment in mice were investigated
using a step-down type passive avoidance task. Eight-week-old male ddY mice were exposed to CO 3
times at 1 hour intervals, 7 days before the first training and retention test and 24 hr after the first training
session. Memory deficiency occurred in mice when training commenced 7 days after CO exposure. The
median step-down latency in the retention test of the CO-exposed group was significantly shorter than that
of the control group. Administration of (-)-nicotine (7.8 - 250 nmol/kg, i.p.) 15 min before the first
training session prolonged the step-down latency in the CO-exposed group, producing a bell-shaped curve.
The effect of (-)-nicotine (15.6 and 31.3 nmol/kg) was significant. To determine whether this effect of (-)-
nicotine was mediated via nicotinic cholinergic receptors, we attempted to block its action using a nicotinic
acetylcholine receptor antagonist (mecamylamine). Mecamylamine (1.25 μmol/kg) blocked the effect of (-)-
nicotine (31.3 nmol/kg) on learning impairment. Administration of (-)-nicotine (15.6 - 62.5 nmol/kg)
immediately after the first training session failed to ameliorate learning ability in the CO-exposed group.
The dosages of (-)-nicotine used in the present study were far below that required to alter either the shock
sensitivity or locomotor activity. The contents of dopamine and its metabolites, DOPAC or HVA in the
frontal cortex of CO-exposed group were high compared to the control group. (-)-Nicotine (15.6 and 31.3
nmol/kg) diminished these changes. On the other hand, dopamine turnover, measured by the ratios
DOPAC/dopamine and HVA/dopamine, indicated (-)-nicotine (15.6 and 31.3 nmol/kg) stimulated
dopamine turnover in the striatum of CO-exposed mice. These results suggest that (-)-nicotine potentiates
the nicotinic cholinergic neuronal system and may potentiate acquisition of memory. Modulating effects of
the dopaminergic neurotransmission may also involve in the ameliorating effect by (-)-nicotine in the
impairment of learning ability induced by CO exposure.

EFFECTS OF NICOTINE IN COGNITIVE IMPAIRMENT - A STUDY USING EVENT-
RELATED POTENTIALS AND MIDDLE LATENCY RESPONSE

S. Katayama, K. Hirata, H. Tanaka, K. Yamazaki, M. Fujikane and Y. Ichimaru. Dept. of
Neurology, Dokkyo University School of Medicine, Tochigi, Japan, 321-02.

In order to evaluate the effects of nicotine administration in cognitive impairment, the electrical field
distribution of event-related potentials (ERP's) and middle latency response (MLR) were analyzed.

The study was carried out on 6 normal individuals and 16 patients with dementia(vascular
dementia, Alzheimer disease and Parkinson disease).

For the ERP measurements the auditory oddball stimulation was presented. Eighty dB 1/sec click
stimuli were presented for the MLR measurements. The EEG was recorded from 18 electrodes placed
according to international 10/20 method. In addition to smoking, nicotine was delivered transdermally
from a nicotine patch.

Dementing patients showed abnormal ERP's in latency, amplitude and electrical field on the scalp.
Decreased amplitude and electrical field abnormality of P1 in MLR was also seen in some patients
with dementia. These abnormal ERP's and MLR of the patients improved after administration of the
nicotine especially in N200 of ERP's and P1 of MLR.

These data suggest that nicotine administration might be useful in dementia as a cognitive enhancer.

COGNITIVE AND PSYCHOMOTOR PERFORMANCE EFFECTS OF REPEATED NICOTINE
DOSING IN NONSMOKERS. S.J. Heishman and J.E. Henningfield. Clinical Pharmacology
Branch, Addiction Research Center, National Institute on Drug Abuse, Baltimore, Maryland,
21224, U.S.A.

It is well documented that impaired cognition and performance can accompany tobacco
abstinence in nicotine-dependent persons and that various forms of nicotine delivery can reverse
such withdrawal-induced deficits in several areas of performance. However, evidence that nicotine
reliably enhances performance under conditions of no deprivation (nonabstinent smokers or
nonsmokers) is limited to motor abilities and certain attentional tests. The purpose of this study
was to determine if nicotine reliably enhanced a wider range of performance measures in
nonsmokers. Twelve male volunteers, who reported ever smoking less than five cigarettes, lived
on an inpatient research unit and participated in 9 consecutive experimental days in which they
were administered various doses of nicotine polacrilex gum for 15 minutes four times each day
(0900, 1030, 1300, and 1430). Before and after each dose, cognitive and psychomotor
performance was assessed. On day 1, only placebo was given. On days 2-9, four doses of
polacrilex were administered each day in this order: 0, 2, 4, and 8 mg. Nicotine did not enhance
performance on either a psychomotor or several cognitive tests. Performance on the circular lights
task, a measure of eye-hand coordination, was significantly impaired by nicotine, especially at the
8 mg dose. Nicotine increased rate of responding and decreased response time on three cognitive
tests, including letter searching, logical reasoning, and digit recall; however, accuracy on these
tests was significantly impaired by nicotine. Performance on a serial addition/subtraction test was
not affected. Over 8 consecutive days of dosing, there was no evidence of tolerance developing to
the performance effects of nicotine.

P 97   COMPARISON OF THE EFFECTS OF SMOKING NORMAL NICOTINE AND NO NICOTINE
        CIGARETTES ON COGNITIVE AND PSYCHOPHYSIOLOGICAL PARAMETERS.

B. Baldinger, M. Hasenfratz and K. Bättig.
Swiss Federal Institute of Technology, Behavioral Biology Lab, Schorenstr. 16, CH-8603
Schwerzenbach, Switzerland.

Smokers often report that smoking enables them to better concentrate on daily work, and improvements in cognitive and vigilance performance have been demonstrated in many laboratory experiments. In order to investigate the effect of nicotine on task performance this study compared the subjects' habitual cigarette ($\geq$ 0.6 mg nic./cig.) with a nearly nicotine free test cigarette (0.08 mg nic./cig.) which had a tar yield comparable to that of the habitual brand. Twenty female smokers participated in two sessions, smoking their habitual or the test cigarettes and performing a rapid information processing task with the stimuli presented at a fixed rate in one group and at a subject-paced rate in a second group of 10 subjects. A first 20-min. trial was done before and a second one during smoking, and physiological and subjective parameters were recorded. Pre- to postsmoking increases in electrocortical and cardiovascular arousal were observed only with the habitual but not with the test cigarettes. As the two cigarettes differed only in their nicotine yield (carbon monoxide uptake was similar with both cigarettes) the different effects of the two brands were interpreted as nicotine effects. Whereas nicotine similarly affected physiological parameters in both types of tasks, it affected performance on the two task versions in different ways. With the fixed presentation rate version of RIP, nicotine reduced the reaction time to hits but did not affect hit probability. With the subject-paced presentation rate version of RIP, nicotine increased the individual processing rate but did not affect the reaction time to hits. The subjective ratings of performance increased from the presmoking to the smoking trial in the subject-paced group when smoking the habitual cigarettes and in the fixed rate group it decreased when smoking the test cigarettes.

P 98   COMPARISON OF THE EFFECTS OF PRETASK SMOKING AND SMOKING DURING A
        TASK ON COGNITIVE AND PSYCHOPHYSIOLOGICAL PARAMETERS.

M. Hasenfratz and K. Bättig.
Swiss Federal Institute of Technology, Behavioral Biology Lab, Schorenstr. 16, CH-8603
Schwerzenbach, Switzerland.

Whereas in daily life smokers usually smoke during work, in most laboratory experiments the subjects have to smoke before performing a task. In this study the effects of smoking while performing a task were compared with those of pretask smoking. Rapid information processing (RIP) performance, physiological and subjective parameters were assessed in twenty female regular smokers, who were allowed, according to a 2 x 2 cross-over design, to either real or sham smoke a single habitual cigarette between the two RIP trials of a session and to real or sham smoke ad libitum during the second RIP-trial. RIP performance was significantly better when the smokers were allowed to smoke during the task, whereas the improvement after pretask smoking failed to reach significance. The puffing intervals were similar for real and sham smoking during rest but significantly longer with real than with sham smoking during the task. The increases in electrocortical and cardiovascular arousal as well as the subjective effects were similar for pretask and task smoking. After pretask smoking these effects vanished for the subsequent task period, whereas they were maintained but not further increased when pretask smoking was followed by task smoking. It was concluded that mental performance was increased but overarousal was avoided by accurate nicotine titration during the RIP task.

P 99   COGNITIVE DEFICIT INDUCED BY NICOTINE ABSTINENCE. J. Le Houezec and Karl O. Fagerström.
CNRS EP53, Hôpital de la Salpêtrière, Paris, France and Pharmacia Consumer Pharma, Helsingborg, Sweden.

This study was designed to show the cognitive impairments induced by nicotine abstinence in smokers trying to sustain attention in a low-demanding task condition.

Subjects: 6 dependent smokers (score $\geq$ 6 on Fagerström test) smoking $\geq$ 20 cigarettes per day.

Procedure: Smokers were asked to come on 2 occasions separated by at least 2 days. On one occasion they abstained from smoking from 22:00 the night before (checked by CO measurement). On the second occasion they were free to smoke until testing started (from 14:00 to 17:00). Food and coffee consumption were held constant between sessions. During task, subjects were asked to chew 3 pieces of gum (active or placebo) for 30 min each at 20 min, 80 min, and 140 min after starting. They were instructed on how to chew the gum prior to be included in the study to prevent side effects due to abnormal use of the gum or sickness due to the dose (4 mg).

Task: Colored squares (3x3 cm) or a white dot ($\varnothing$ 3 cm) were presented successively on a video screen, 1.5 m from subjects' eyes, at an average rate of 1 per 18 sec (ISI between 5 sec and 30 sec). Subjects were asked to respond to squares only by pressing the left (red squares) or the right (yellow squares) button of a computer mouse held on their knees. Squares were presented with a probability of .2 (red + yellow = .4), dots were presented with the remaining probability of .6. Reaction time (RT), errors, and latency and amplitude of N1 and P3 components of the Event-Related Potentials (ERP) were recorded.

Results: Preliminary results show that RT was increased in the abstinent condition compared to the non-abstinent condition. The amplitude of the P3 component of the ERP was dramatically reduced in the abstinent condition, reflecting a reduction in cognitive resource allocation. The P3 latency was longer in the abstinent condition reflecting an impairment in stimulus processing. ERP is a precious tool for assessing cognitive impairments induced by nicotine abstinence.

AN INTRA-LABORATORY CONTRAST OF THE COGNITIVE AND PSYCHOMOTOR EFFECTS
OF NICOTINE, ALCOHOL AND DOTHIEPIN.

N. Sherwood and I. Hindmarch. Human Psychopharmacology Research Unit, University of Surrey,
Milford Hospital, Godalming, GU7 1UF, United Kingdom.

Recent advances in effect-size and meta-analytic techniques have suggested that it may be possible to
combine and contrast the results of separate experiments, conducted using the same measures and
methods, within a statistical framework and to quantify the relative influence of each experimental
parameter. A pilot study was conducted incorporating the results of three HPRU studies into the
behavioural effects of nicotine, alcohol and dothiepin (a tricyclic antidepressant) in healthy young
volunteers. Each study was treated as a between-groups factor in an exploratory analysis of variance
which, in addition to treatment and time of testing included terms such as chronological date of study,
number of subjects, subject sex and investigator status. In lieu of generating F values, contrast analyses
were conducted on sums of squares for each ANOVA term. These showed that the measures and methods
were consistent in their sensitivity to drug effects and moreover that the positive effects of nicotine were
clearly delineated from alcohol and dothiepin, both of which were associated with increased subjective
sedation and impaired cognitive and psychomotor function. Differences in placebo responses between each
experiment suggested that expectancy effects may influence such results. However, when these differences
were controlled, the relative influence of the three compounds was effectively unchanged. These results
highlight certain inconsistencies in the perceived psychological benefits and actual behavioural risks
associated with such compounds and their use in everyday life.

Section 12: Nicotine psychopharmacology in animals

DIFFERENT METHODS OF ASSESSING NICOTINE-INDUCED ANTINOCICEPTION
MAY ENGAGE SOMEWHAT DIFFERENT NEURAL MECHANISMS A. R. Caggiula, L.
H. Epstein, K. A. Perkins, and S. Saylor Departments of Psychology and Psychiatry,
University of Pittsburgh, Pittsburgh, PA 15260 and Pittsburgh Cancer Institute

Two methods were used to assess nicotine-induced antinociception: tail-
withdrawal from a hot water bath and hind paw withdrawal from a hot plate. Doses
of nicotine which produced 75-80% maximum response were 0.75 mg/kg (free base)
for tail-withdrawal and 0.35 mg/kg for hot-plate. The peripheral blocker
chlorisondamine (CHLOR; 0.1 mg/kg, sc) was as effective as the central/peripheral
antagonist, mecamylamine (MEC; 1 mg/kg, sc) in blocking nicotine-induced
antinociception as measured by the tail-withdrawal method, suggesting that this
measure depends on either the action of nicotine at peripheral receptors or the
functional integrity of those receptors (e.g., uninterrupted impulse flow at
sympathetic ganglia). In contrast, nicotine-induced antinociception, as measured
by the hot-plate method, was blocked by MEC but not CHLOR. These results
indicate that the two methods of assessing nicotine-antinociception involve at least
partially separate pathways and cannot be used interchangeably. In an initial study
to further determine the pathways mediating nicotine's antinociceptive effects, rats
were pretreated with MEC or the β-adrenergic blocker, propranolol and tested using
the tail-dip paradigm. As before, MEC completely blocked nicotine-antinociception
but propranolol was without effect. Supported by DA07546

P 102

ANTINOCICEPTIVE EFFECTS OF CENTRAL AND PERIPHERAL ADMINISTRATION OF NICOTINE IN THE PRESENCE OF FORMALIN PAIN IN RATS. A. A. Houdi and M. Welch. Tobacco and Health Research Institute and College of Pharmacy, University of Kentucky, Lexington, KY 40546.

Previous studies indicate that a single dose of nicotine administered systemically or centrally produced an analgesic effect as measured by tail-flick technique. These findings, together with previous failure to obtain nicotine-induced analgesia in other pain tests, raises the possibility that nicotine may selectively alter sensitivity only to certain classes of pain stimuli. In this study, we examined further the central and peripheral antinociceptive properties of nicotine in rats using formalin-test. It has been suggested that the transient early phase of pain (0-5 min) after subcutaneous (sc) formalin is due to the direct effect on sensory receptors, whereas the tonic late phase (15-45 min) is due to an inflammatory response. In control groups, formalin solution (0.5%) injected subcutaneously into the surface of one hind paw of male Sprague-Dawley rats (300 ± 30 gm, body wt.) pretreated with saline produces a characteristic biphasic pain response. Subcutaneous administration of nicotine inhibited this formalin response in a dose-dependent (0.1-2 mg/kg) manner. Intracerebroventricularly administered nicotine inhibited the response of the first phase at lower dose (4 µg/rat) and the first phase and partially the second phase at higher doses (20 & 40 µg/rat). Nicotinic antagonist mecamylamine (0.5 mg/kg, sc) antagonized the sc nicotine effect, whereas hexamethonium (5 mg/kg, sc), a nicotinic antagonist which does not readily cross the blood brain barrier, partially blocked the response only to the second phase of nicotine's effect. Pretreatment with an opioid receptor antagonist (naloxone 2 mg/kg, sc), or nitric oxide synthetase inhibitor (N^G-nitro-L-arginine methyl ester, 50 mg/kg, sc) had no effect on the analgesic properties of subcutaneous nicotine (1 mg/kg). These results indicate that nicotine possesses analgesic activity toward two kinds of pain which may be differently modulated; a short-lived pain caused by the direct effect of formalin on sensory receptors followed by a longer lasting pain due to inflammation. This analgesic effect of nicotine is not mediated by endogenous opioids or nitric oxide systems. Supported by KTBR.

P 103

ANTINOCICEPTIVE EFFECT OF CHRONIC NICOTINE IS POTENTIATED BY NIFEDIPINE. V.K. Zbuzek, T. Glasser and W. Wu. Department of Anesthesiology, University of Medicine & Dentistry of New Jersey, New Jersey Medical School, Newark, NJ 07103.

Recently, we brought evidence that pretreatment with the calcium channel blocker nifedipine (NIF) significantly potentiated antinociceptive effect of a single dose of nicotine (NIC) (1). The present study was designed to investigate the extent to which the antinociceptive effect of chronic NIC, which was detectable only by hot-plate but not the tail-flick technique, (2) is altered by a single dose of NIF. NIC (6 mg/kg/day) was administered s.c. via Alzet osmotic pumps to rats for 28 days and the nociception was measured by the tail-flick method. NIF (15 mg/kg i.p.) was injected to different rats, during chronic nicotine infusion, at day 3 and 21, and 2 weeks after NIC withdrawal. In control (sham operated) rats NIF exhibited the maximal median latency 30 min after the injection, and lasting for 5 min. In rats chronically infused with NIC, NIF revealed the maximal latency within 10 or 25 min and lasting for 70 or 60 min at day 3 or 21 on NIC, respectively. NIC withdrawal inhibited this phenomenon. These data support our hypothesis that calcium is involved, at least in part, in nicotine-induced antinociception. It further suggests that chronic smokers who are being treated with NIF could be at potential risk to develop an increased pain threshold and miss a first sign of myocardial ischemia, chest pain.

1) Wong et al, Life Sci (in press); 2) Yang et al, Psychopharmacol 106, 417, 1992

P 104

EFFECTS OF ACUTE NICOTINIC BLOCKADE ON THE ACOUSTIC STARTLE RESPONSE AND PRE-PULSE INHIBITION IN MALE AND FEMALE RATS. E. J. Popke and N. E. Grunberg. Dept. of Medical and Clinical Psychology, Uniformed Services University of the Health Sciences, Bethesda, Md, USA, 20814-4799.

Nicotine has an inverted U-shaped dose-effect on the acoustic startle response (ASR) and pre-pulse inhibition (PPI) in male and female rats. Because ASR and PPI are thought to be indices of attentional and sensory-gating mechanisms, these measures provide a useful paradigm to evaluate attentional effects of nicotine. The mechanisms by which nicotine exerts these effects, however, have not been identified. The present experiments examined the effects of peripheral s.c. administration of mecamylamine (1.0 mg/kg and 2.0 mg/kg) and hexamethonium (1.0 mg/kg and 2.0 mg/kg) on ASR and PPI of male and female Sprague-Dawley rats. The effects of both antagonists were similar to those of nicotine in that moderate doses increased ASR and PPI, whereas higher doses returned ASR and PPI to control levels. Implications of these data for the mechanisms underlying attentional effects of nicotine will be discussed.

EFFECTS OF NICOTINE AND METHYLPHENIDATE ON RESPONDING MAINTAINED
BY EITHER COCAINE OR FOOD REINFORCEMENT. Domenico Meloni, Tim Koves
John Robinson and Steven I. Dworkin. Department of Physiology and Pharmacology,
Bowman Gray School of Medicine, Wake Forest University, Winston-Salem, NC 27157,
USA.

Attempts to demonstrate the reinforcing effects of nicotine using self-administration
procedures have resulted in equivocal findings. The present study evaluated the
effects of nicotine (0.1 - 1.7 mg/kg, i.p.) and methylphenidate (0.3 - 30.0 mg/kg) on
responding maintained by cocaine or food presentation in male Fisher rats. Both the
food and cocaine reinforcers (0.17 and 0.33 mg/inf) were presented under a fixed-ratio
10 schedule. The food schedule also included a 6 min time-out following each food
presentation to generate response patterns similar to those resulting from cocaine
deliveries. Methylphenidate, which will engender and maintain self-administration,
resulted in dose related decreases in responding maintained by food and cocaine.
However, nicotine did not alter responding maintained by either food or cocaine.
Previous studies have shown that pretreatment with cocaine or heroin decreased
responding maintained by cocaine. Thus, drugs with high abuse potential
(unequivocally supporting self-administration) appear to modulate the self-
administration of cocaine. The results from this study suggest that methylphenidate
(which is reliably self-administered) can alter the reinforcing effects of cocaine, while
nicotine (which is equivocal in self-administration studies) does not influence cocaine
self-administration. (Research supported by a gift from the RJR Tobacco Co.)

THE EFFECTS OF NICOTINE PRETREATMENT ON LOCOMOTOR RESPONSES TO d-
AMPHETAMINE   C.E. Birrell and D.J.K. Balfour   Department of Pharmacology, University
Medical School, Ninewells Hospital, Dundee Scotland DD1 9SY

Repetitive daily injections of nicotine (0.4mg/kg sc) for five days enhances the locomotor (LMA)
and mesolimbic dopamine (DA) responses to the drug [1]. In this study the influence of the same
pretreatment schedule on responses to d-amphetamine have been examined. On the test day groups
of saline- or nicotine-pretreated rats were habituated to a 4-arm maze for 1 hour before being given
sc injections of saline, nicotine (0.4mg/kg) or d-amphetamine (0.2mg/kg). Pretreatment with
nicotine enhanced the peak response to nicotine from 15 ± 3 to 34 ± 5 and d-amphetamine (P<0.05)
from 18 ± 2 to 30 ± 5 entries per 20 minutes. A separate study using an activity box also revealed an
enhanced response (P<0.05) to d-amphetamine from 76 ± 33 to 209 ± 39 beam breaks per 20 minutes.
This effect appeared to be associated with increased DA overflow in the nucleus accumbens
although this effect did not reach significance. The results suggest that nicotine pretreatment
sensitises rats to the LMA response to d-amphetamine when they are tested in animals which have
been habituated to the test environment and that the cross-sensitisation may be associated with
increased DA overflow in the nucleus accumbens although this requires confirmation.

This study was supported by a grant from the Wellcome Trust

1. Benwell MEM & Balfour DJK (1992) Br J Pharmacol, 105: 849-856.

CORRELATIONS BETWEEN NICOTINE-INDUCED LOCOMOTOR ACTIVITY AND ETHANOL
PREFERENCE IN WISTAR RATS. O. Blomqvist, D. Johnson, J.A. Engel and B. Söderpalm. Dept. of
Pharmacology, Göteborg University, Medicinaregatan 7, 413 90 Göteborg, Sweden.

Co-abuse of ethanol and nicotine is frequent in man, possibly explained by a common mechanism of
action of these drugs. We have recently presented data suggesting that ethanol, like nicotine, activates the
mesolimbocortical DA system through a direct interaction with central nicotinic acetylcholine receptors, and
that nicotine sensitization may increase ethanol preference and intake in Wistar rats. The aim for the present
study in rats was to investigate 1) if sensitivity to the locomotor stimulating effect of nicotine could predict
future ethanol preference, and 2) if long-term ethanol intake altered the sensitivity to nicotine. Drug naive
Wistar rats were challenged with nicotine (0.4 mg/kg s.c.), and locomotor activity was measured for 60
minutes. Three weeks later, an ethanol solution was presented to the animals, and after six weeks in a free
choice drinking-situation (ethanol 6% v/v or water) they were classified as low-(LP; <20% ethanol),
medium-(MP; 30-60% ethanol) or high-(HP; >70% ethanol)preferring based on their ethanol preference.
After 3 weeks without access to the ethanol solution, locomotor activity after nicotine challenge was
measured again. There was no clear correlation between locomotor stimulation after the first nicotine
challenge and subsequent ethanol preference, but rats that were later found in the HP group was initially
poorly stimulated by nicotine. Nicotine-induced locomotor activity was significantly higher at the second
challenge compared to the first, when measured in all animals together, and at this point HP rats were
significantly more stimulated after nicotine than LP rats. There was a positive correlation between ethanol
preference and the relative increase in nicotine stimulation between the two challenge occasions. These data
suggest that rats, that might develop high preference for ethanol display an initial low sensitivity to the
locomotor stimulating effect of nicotine, and that long-term ethanol consumption might cross-sensitize rats
to the locomotor stimulating effect of nicotine.

ANTIAMNESTIC EFFECT OF NICOTINE IN MOUSE PASSIVE-AVOIDANCE TEST

C. Ghelardini, N. Galeotti, P. Malmberg-Aiello, A. Giotti and A. Bartolini Dept. of Pharmacology, University of Florence, Viale G.B. Morgagni, 65, I-50134, Florence, Italy.

Much evidence suggest the ability of nicotine, via a positive feed-back mechanism, to increase ACh release both in the CNS and in the periphery through stimulation of presynaptically localized nicotinic receptors (Beani et al., 1985; Bowman et al., 1988; Rowen and Winkler, 1984). Iwamoto (1989) described the possibility to prevent ACh releasing action by pretreatment with (-)-vesamicol, an agent that interferes with storage/or release of ACh. ACh is considered a primary neurotransmitter in mnemonic function; therefore the purpose of this research was to evaluate the role of nicotine in learning and memory processing. Nicotine was tested in the mouse in a modified one trial pass-through passive avoidance test consisting of a painless punishment (fall into cold water, 10°C). Mouse cognitive processes had been impaired by dicyclomine (2 mg/kg i.p.) or scopolamine (1 mg/kg i.p.) both administered immediately after the training session. In this test nicotine at the dose 0.1-0.5 mg/kg s.c. and 0.3-1μg/mouse i.c.v., injected 10 min before the training session was able to protect animals from amnesia induced by antimuscarinic treatments. In the same range of doses nicotine was unable to improve learning in mice devoid of memory impairment. Effective doses of nicotine failed to affect rota-rod performance and spontaneous locomotor activity. This study supports the possibility of prevening cognitive deficits through activation of nicotinic autoreceptors by using nicotine at lower doses than those responsible for a post-synaptic effect. It could represent a novel therapeuthic strategy for the treatment of dementia.
This work was supported by grants from MURST and CNR

LACK OF TOLERANCE TO THE DISCRIMINATIVE STIMULUS EFFECTS OF NICOTINE IN RATS.

Mohammed Shoaib, Eric Thorndike, Steven R. Goldberg & Charles W. Schindler. Preclinical Pharmacology Branch, Addiction Research Center, National Institute on Drug Abuse, Baltimore MD 21224, U.S.A.

Tolerance to discriminative stimulus (DS) effects of drugs, as observed by a shift of the dose-response curve to the right, has been observed with many addictive drugs (e.g. amphetamine, cocaine and morphine). Chronic administration of nicotine has been reported to produce tolerance to the locomotor depressant effects and aversive stimulus properties of nicotine. However, the DS effects of nicotine have not been examined for development of tolerance following chronic treatment. We report on experiments utilising a discrimination paradigm as previously described by Young et al. (*Pharmac Biochem Behav* 39: 487, 1991). Eight, male Sprague-Dawley rats were trained to discriminate nicotine (0.4 mg/kg s.c.) from saline under a fixed ratio (FR 10) schedule for food reinforcement. Multiple training sessions were given daily, and once criteria was met, cumulative doses of nicotine (0.025-1.2 mg/kg s.c.) were evaluated. Rats acquired the nicotine discrimination after 80 sessions. During this period, rats developed tolerance to the rate-depressing effects of nicotine after 20 nicotine training sessions. In rats that showed stable dose-related responding, daily injections of saline for 7 days in the rat's home cage during suspended training failed to shift the dose-response curve. A similar regimen of chronic treatments with the training dose of nicotine (0.4 mg/kg s.c.) and with a larger dose of nicotine (1.2 mg/kg s.c.) failed to shift the dose-response curve. These results suggest that rats trained to discriminate 0.4 mg/kg nicotine from saline do not readily develop tolerance to the DS effects of nicotine.

NICOTINE PATCH DOSE EFFECTS ON SLEEP: SELF-REPORTED OUTCOMES AND
ASSOCIATED SALIVA NICOTINE LEVELS

Scott J. Leischow, Ph.D., Suzanne N. Valente, M.D., Anabel L. Hill, M.S., Pamela S. Otte, B.S., Mikel
Aickin, Ph.D., Evan W. Kligman, M.D.
The University of Arizona, Smoking Research and Treatment Program, 1233 N. Santa Rita, Tucson,
AZ 85719

The present within-subjects, crossover study was designed to assess the short-term effects of several
nicotine replacement strategies. After completing a 2 day smoking baseline, 18 (7 M, 11 F) smokers
($\geq$ 20 cpd) went through the following conditions, each for two full days, in randomized order: a)
nicotine chewing gum; b) 24 hour nicotine patch; c) 16 hour nicotine patch; d) two 24 hour nicotine
patches; e) 24 hour nicotine patch plus nicotine gum; f) no medication. Only conditions b, c, d, and f
will be compared here. Subjects provided self-report and objective data on 5 occasions during each 2
day period (7-8 AM & 5-6 PM), and smoked ad lib for the 5 days between conditions. Between group
comparisons of **abnormal dreams** found (at AM visit): 24 hour patch & double 24 hour patch >
baseline. Between group comparisons of **interrupted sleep** found (at AM visit): double 24 hour patch
> baseline. Between group comparisons of **difficulty getting to sleep** found (at AM visit): double 24
hour patch > baseline & 16 hour patch. Saliva nicotine and cotinine values will be presented. These
results suggest that use of a double strength nicotine patch can have significant effects on sleep. While
some subjects on 24 hour patch reported sleep disturbances, there was no significant difference
between 16 & 24 patch use on self-reported sleep disturbances.

DISCRIMINATIVE STIMULUS EFFECTS OF NICOTINE IN SMOKERS. K. Perkins, J. Grobe,
A. Scierka, R. Stiller. Depts. of Psychiatry & Anesthesiology, Univ. of Pittsburgh School of
Medicine, Pittsburgh, PA 15213, USA.

Interoceptive stimulus effects may be critical in determining reinforcing efficacy of abused drugs,
including nicotine. Although traditional subjective measures may be useful, another, perhaps more
reliable, method of gauging these stimulus effects is the drug discrimination procedure. Animals have
been shown to reliably discriminate among nicotine doses and between nicotine and other drugs, but
there has been virtually no research examining discrimination of nicotine per se in humans using
formal drug discrimination procedures. In this study, male and female smokers (n=9 each) were
trained on Day 1 to reliably discriminate 0 vs. 12 ug/kg nicotine administered by measured-dose nasal
spray. All Ss were able to reach criterion performance (at least 80%
correct). Generalization of responding across 0, 2, 4, 8, and 12 ug/kg
nicotine doses (0-0.8 mg for typical S) was then examined on Day 2.
Nicotine-appropriate responding was linearly related to dose (see
figure), and subjects were able to distinguish the smallest dose (2
ug/kg) from placebo. Although there were no differences between
males and females in behavioral discrimination, subjective effects were
correlated with nicotine discrimination in females but not in males.
These findings indicate that humans are able to discriminate among
low doses of nicotine per se, that males and females may differ in the
stimuli used to discriminate nicotine, and that drug discrimination
procedures may be more sensitive than traditional subjective effects
measures in distinguishing among low doses of nicotine.

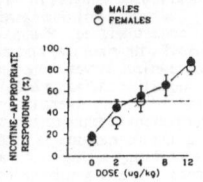

Supported by National Institute on Drug Abuse Grant DA-08578 (KAP)

THE EFFECTS OF NICOTINE ON PERCEPTUAL SPEED. C.Stough, G.
Mangan, and T. Bates. Department of Psychology, University of Auckland, Private
Bag 92019, Auckland, New Zealand.

Previous studies have reported an enhancement in information processing speed in
nicotine conditions compared to no-nicotine conditions, employing RT procedures.
However, these procedures do not allow a distinction to be made between early or late
processes of information processing. In order to examine whether this nicotine
related enhancement in information processing is due to early or later processes we
conducted an experiment investigating the effects of nicotine on inspection time
(IT). IT has been employed in several paradigms and reflects the early processes of
information processing relating to the registration and integration of stimulus
information. IT was administered to 50 subjects in non-smoking, sham-smoking, .4,
.8 and 1.2 mg nicotine conditions. Results suggest that nicotine enhances the early
processes of information processing. This result supports converging evidence for a
role of the cholinergic system in intellectual performance. A cholinergic model of
intelligence is discussed.

THE SENSORY ROLE OF NICOTINE IN CIGARETTE 'TASTE', SMOKING SATISFACTION, AND DESIRE TO SMOKE. Walter S. Pritchard and John H. Robinson. Psychophysiology Laboratory, Bowman Gray Technical Center, R. J. Reynolds Tobacco Company, Winston-Salem NC 27102.

Thirty-two subjects were tested in five double-blind sessions (16 in the morning following overnight smoking abstention, and 16 in the afternoon following ad-lib smoking). In each session, subjects smoked one of five experimental (EX) cigarettes, with the order of cigarettes within sessions balanced across subjects. The EX cigarettes had the following nicotine/'tar' yields in mg: 0.08/8.5, 0.17/9.1, 0.37/9.8, 0.48/9.8, and 0.74/10.4. In a sixth session, subjects smoked a 0.71/8.6 commercial 'light' (CL) cigarette. Before and after smoking, subjects subjectively rated their desire to smoke (DTS) a cigarette of their usual brand and had blood samples drawn. Following smoking, subjects rated the cigarette on a variety of sensory dimensions; they also rated how satisfying the cigarette was. Nicotine yield had a significant impact on a variety of sensory dimensions (perceived 'draw' [ease of getting smoke through filter]; 'strong' taste, 'tobacco' taste; nose impact; chest impact). The CL cigarette was rated much easier to 'draw' and to have more chest impact than the EX cigarettes. On a pre- to post-smoking basis, the increase in blood nicotine was largely a linear function of nicotine yield. For nicotine yields >= 0.48, the increase in blood nicotine was smaller in A.M. subjects than in P.M. subjects. ΔDTS (DTS pre minus DTS post) also tracked nicotine yield in a largely linear manner, with the 0.71 CL and the 0.74 EX cigarette producing virtually identical ΔDTS's – thus, ΔDTS appeared to be largely a function of nicotine pharmacology. However, for satisfaction, the following pattern was obtained: 0.08 = 0.17 < 0.37 = 0.48 = 0.74 < 0.71 (CL) – thus, satisfaction did *not* track nicotine yield in a linear manner, and the better 'draw' and increased chest impact of the CL cigarette appeared to result in an additional increase in smoking satisfaction. It is concluded that [1] nicotine plays an important role in the 'taste' of cigarettes, and [2] non-pharmacological factors play an important role in smoking satisfaction.

CHRONIC AND ACUTE TOLERANCE TO NICOTINE'S SUBJECTIVE, BEHAVIORAL, AND CARDIOVASCULAR EFFECTS IN HUMANS. K. Perkins, J.Grobe, C. Fonte, A. Caggiula, A. Scierka, R. Stiller. Departments of Psychiatry, Psychology, and Anesthesiology, Univ. of Pittsburgh School of Medicine, Pittsburgh, PA 15213, USA.

Understanding tolerance to nicotine in humans may clarify processes responsible for the onset and maintenance of tobacco dependence. Tolerance to nicotine has been clearly demonstrated in animals, but human research is lacking. Effects of nicotine on subjective (POMS, visual analog scales), behavioral (psychomotor and cognitive tasks), and cardiovascular (HR, SBP, DBP) responses were examined as a function of past history of nicotine exposure (i.e. smokers vs. nonsmokers, chronic tolerance) and of immediately preceding nicotine exposure (acute tolerance). Dose-effect relationships between nicotine (0-20 ug/kg via measured-dose nasal spray) and each response were determined in male and female smokers (n=17) and nonsmokers (n=18), with different doses presented on different days. Following this dosing regimen on each day, Ss also received a challenge dose of 20 ug/kg to assess acute tolerance. Plasma nicotine concentrations were unexpectedly 30% lower in nonsmokers compared with smokers, and analyses were adjusted to control for this difference. Results showed that dose-effect curves were shifted to the right or dampened in smokers relative to nonsmokers for most subjective and some behavioral responses, consistent with chronic tolerance, but there was less evidence of chronic tolerance to other behavioral effects or to cardiovascular responses. A generally similar pattern of acute tolerance was observed across responses. These results provide evidence that regular use of nicotine is associated with chronic functional tolerance and that repeated nicotine exposure during a single episode produces acute tolerance. A similar pattern of chronic vs. acute tolerance suggests similarity of mechanisms responsible for both "types" of tolerance. However, variability in tolerance magnitude across subjective, behavioral, and cardiovascular response domains indicates that different mechanisms may be responsible for these different effects of nicotine. *Supported by National Institute on Drug Abuse Grant DA-05807 (KAP)*

EEG FREQUENCY CHANGES FOLLOWING TOBACCO SMOKING IN RELATION TO PLASMA NICOTINE LEVELS. E.F. Domino, C. Kadoya, and S. Matsuoka, Department of Pharmacology, University of Michigan, Ann Arbor, MI 48109-0626 USA and Department of Neurosurgery, University of Occupational and Environmental Health, Kitakyushu 807, Japan

Electroencephalographic (EEG) changes from scalp electrodes in chronic tobacco smokers have been described by many investigators including Murphree and Schultz (1968); Ulett and Itil (1969); Itil *et al*. (1971); Knott and his colleagues (1997, 1988, 1991, 1994); Herning *et al*. (1983); Pickworth *et al*. (1989); and Domino *et al*. (1992, 1994). Except for the studies by Knott *et al*. and Domino *et al*., most of the previous studies did not involve multichannel topographic EEG mapping because of technical limitations in computer hardware and software. Surprisingly detailed measurements of plasma nicotine levels during tobacco abstinence and their relationship to EEG effects have been lacking. The purpose of this research was to determine the change in the plasma level of nicotine that would affect the baseline EEG of tobacco smokers deprived of tobacco overnight. Sixteen channels of monopolar EEG scalp recordings were correlated with plasma nicotine and cotinine levels. The latter were measured by HPLC techniques. Nonsmokers inhaled air through a placebo cigarette to mimic smoking. Following air inhalation, nonsmokers showed relatively minor EEG changes. In contrast, tobacco smokers showed important EEG effects. Mean *delta* and *alpha*$_1$ activities were significantly decreased and *beta* activity was increased by increasing plasma nicotine concentrations. The tobacco smokers had a statistically significantly higher dominant *alpha* frequency following smoking compared to the nonsmokers (P<.01). The dominant *alpha* frequency increased significantly with increasing plasma nicotine levels. Increases of plasma nicotine greater than 10 ng/ml produced the most obvious EEG effects. (Supported in part by NIDA grant DA-07226 to E.F.D.)

THE INFLUENCE OF ORALLY RESORBED NICOTINE ON SMOKING BEHAVIOR. C.Conze, G.Scherer, A.R.Tricker and F.Adlkofer. Analytisch-biologisches Forschungslabor Prof. F.Adlkofer, Goethestrasse 20, 80336 Munich, Germany.

Controversial and conflicting results have been reported from numerous nicotine preload and supplementation studies performed to test the nicotine uptake regulation hypothesis for smokers. We have carried out two cross-over studies in which smokers received either oral nicotine or a placebo according to a fixed schedule under blind conditions over two consequtive days. In Study 1, 5 smokers received 10 oral nicotine (4 mg) applications or placebo and were asked to smoke 1 cigarette every 20 min over 8 h. When receiving nicotine, the subjects smoked less as indicated by higher plasma nicotine concentrations, lower CO levels in blood and exhalate, and leaving longer cigarette butt lengths. However, none of the differences were significantly different to when receiving the placebo. In Study 2, 11 smokers were allowed to smoke ad libitum during three 2 h periods. Each smoker received 10 oral doses of nicotine (4 mg) or placebo both prior to and during the smoking periods. The concentrations of nicotine and cotinine in plasma, and cotinine in saliva were significantly higher in smokers receiving nicotine compared to the placebo. No differences were found for COHb, CO in exhalate, cigarette consumption and butt length. Their are three possible explanations for this result: (i) the time for learning to adjust the way of smoking to the increased nicotine intake was not long enough, (ii) nicotine is not the primary factor governing smoking behavior, or (iii) the study design has resulted in tolerance to nicotine thus abolishing its governing effect during smoking. We consider the latter explanation most probable.

EFFICACY OF NASAL NICOTINE AND NICOTINE INHALERS: TWO PLACEBO-CONTROLLED TRIALS

Nina G. Schneider, Ph.D.

UCLA School of Medicine
VA Medical Center, Brentwood Division
Los Angeles, California

Two nicotine reduction treatments were tested in separate placebo-controlled trials. A nasal nicotine spray (NNS) was compared with a pepper (piperine) placebo in 255 subjects. A nicotine inhaler (puffed) was compared with a menthol placebo in 232 subjects. The NNS is a relatively fast-acting, easily self-administered form of nicotine. The nicotine inhaler is harder to use (extract nicotine from) but allows some secondary reinforcement (oral, handling) with nicotine delivery. Materials presented will include comparative nicotine plasma levels (previously established) and outcome rates for the two trials. Significant differences in success rates using conservative survival analyses (no slips allowed and $CO \leq 8$) were observed for NNS compared to placebo controls. Data will be presented for first week (daily) abstinence and for test intervals at 2 weeks, 3 weeks, 6 weeks, 3 months, 6 months and one year. Additional analyses with slips will show success rates that are higher for both active and placebo subjects. For the nicotine inhaler, conservative (no slip and $CO \leq 8$) outcome data will be presented for test intervals at one week, 2 weeks, 3 weeks, 6 weeks, 3 months, 6 months and one year. These preparations may allow for matching smokers to treatment by speed and route of nicotine delivery and as a function of sensory/ritual needs.

CO-FACTORS FOR SMOKING. Cynthia S. Pomerleau and Ovide F. Pomerleau, Behavioral Medicine Program, Dept. of Psychiatry, Univ. of MI, Ann Arbor MI 48108 USA

In the U.S., a combination of increasingly restrictive regulations and intensive public education campaigns over the past three decades have brought about a dramatic reduction in smoking rates. Despite optimistic hopes for a smoke-free America by the 21st century, however, it could be argued that smoking rate will asymptote at perhaps 20%, and that the American experience may serve as an example not only of the success but also of the limitations of attempts to eradicate smoking. As the "easy quits" and casual smokers are eliminated from the smoker pool or discouraged from initiation via these measures, who will remain in the ranks of smokers? For one thing, they will tend to be highly dependent. This contention is supported by Fagerstrom's observation that the mean nicotine dependence score of American smokers substantially exceeds that of smokers in countries like France, where the current smoking rate is much higher than in the U.S. It is probably also safe to predict that smoking will become more and more concentrated in special populations who, because of various risk factors, are more likely to be recruited to smoking or who may find it more difficult to quit. Co-factors known or suspected to be associated with smoking include depression, eating disorders, and anxiety disorders, in both clinical and subclinical forms. There is considerable evidence to suggest that when such smokers attempt to quit, symptomatology may emerge or increase. It is therefore unlikely that the prognosis for smoking cessation will improve unless appropriate behavioral and/or pharmacological interventions can be developed to address the special needs of smokers with these co-diagnoses; or alternatively, unless nicotine (administered via a relatively safe vehicle) is recognized as a legitimate therapeutic intervention for one or more of these indications. Data from our own work and that of others will be presented.

P 119

AMELIORATION OF PARKINSON'S DISEASE SYMPTOMS BY NICOTINE: DEMONSTRA-TION USING A WITHIN-SUBJECT REVERSAL DESIGN. O.F. Pomerleau, B. Giordani, and F. Stelson, Behavioral Medicine Program, Dept. of Psychiatry, Univ. of MI, Ann Arbor MI 48108 USA

Encouraging results in 13 cases of post-encephalitic Parkinsonism treated with nicotine were published in 1926, but there have been no subsequent reports on the use of nicotine for the clinical management of Parkinsons's Disease (PD). We treated a 64-year-old male PD patient (G.B.) with a combination of acute and chronic nicotine dosing. G.B., a former smoker, exhibited characteristic features of PD, with bradykinesic symptoms predominating. To avoid loss of movement later in the day, G.B. typically slept 4-5 hrs each afternoon (added to 5-6 hrs' sleep at night). Nicotine treatment was initiated using 16-hr nicotine patches, supplemented a month later by nicotine polacrilex. A double-blind, placebo-controlled design was used: *Condition A (baseline), 1 wk*--no nicotine; *Condition B, 11 wks*--ascending dosing with patch reaching 20mg plus 6 pieces of 4mg gum; *Condition C (high dose nicotine), 4 wks*--20mg patch plus up to 11 pieces of 4mg gum; *Condition D (nicotine washout), 2 wks*--placebo patch/placebo gum; and *Condition E (re-dosing), 10 wks*--ascending dosing with patch reaching 35mg, plus up to 4 pieces of 4mg gum. In *Condition C*, G.B. reported more efficient sleep, more energy (afternoon nap down to 1-2 hrs or skipped), and improved mood; as corroborated by both the referring neurologist and G.B.'s wife, speech was clearer and handwriting and facial expression of emotion were improved compared with *Condition A*. Redosing in *Condition E* restored function lost in *Condition D*. G.B.'s self-rating of overall function was proportional to cotinine level, used as an index of nicotine exposure. Though the mechanism for these effects is not fully determined, nicotine modulates dopamine (DA) activity through excitatory cholineceptors on DA cells in the substantia nigra, and repeated, intermittent administration leads to potentiation of meso-limbic DA secretion along with enhanced locomotor responsivity to subsequent dosing in animals; chronic nicotine dosing protects against degeneration of central DA neurons induced by neurotoxic and mechanical lesions. Nicotine therapy for PD is promising and merits further testing.

P 120

SMOKING AND IDIOPATHIC PARKINSON'S DISEASE: A META-ANALYSIS. W.-D.Heller, A.R.Tricker and F.Adlkofer. Analytisch-biologisches Forschungslabor Prof. F.Adlkofer, Goethestrasse 20, 80336 Munich, Germany.

Epidemiological evidence indicates that subjects with idiopathic Parkinson's disease (IPD) are less likely to be smokers than nonsmokers. The present investigation is a first attempt to assess the relationship between smoking and IPD using metaanalytical techniques. The following results have been obtained:

Study group (number of IPD-Cases)		Risk-Ratio estimate		
		Range	Overall est.	95%-Conf.Int.
5 Cohort Studies (N ≈ 350)		0.36 - 0.76	0.55	[0.51,0.60]
24 Case-Con-trol-Studies (N = 3650)	Current smoking	0.12 - 0.72	0.38	[0.31,0.47]
	Ever smoking	0.22 - 1.90	0.64	[0.59,0.71]
	Exsmoking	0.44 - 0.81	0.92	[0.74,1.13]

While the inverse correlation between ever and current smoking and IPD is confirmed in our metaanalyses, the relationship between exsmoking and IPD shows a high variation from study to study. Whether there is a causal relationship - most probably due to nicotine uptake - or only an indirect one caused by concurrent factors remains to be elucidated.

P 121

PITUITARY AND ADRENAL HORMONE RESPONSES TO PHARMACOLOGICAL, PHYSICAL, AND PSYCHOLOGICAL STIMULATION IN HABITUAL SMOKERS AND NONSMOKERS C. Kirschbaum, G. Scherer and C.J. Strasburger. Center for Psychobiology and Psychosomatic Research, University of Trier.

Hormone responses to injection of corticotropin-releasing hormone, following bicycle ergometry, and psychological stress were studied in 10 habitual smokers and 10 nonsmokers. Compared to injection of saline, significant increases were found in ACTH, prolactin, growth hormone, total serum cortisol, and salivary cortisol under all three stimulations, except for salivary cortisol under ergometry. Furthermore, the smokers showed significant elevations of all five hormones investigated following smoking of two cigarettes of the subject's preferred brand. Comparisons of hormone responses between smokers and nonsmokers revealed a general trend towards stronger responses in nonsmokers. The most significant differences between smokers and nonsmokers were evident in their free cortisol and growth hormone responses after psychological stress and ergometry. Moreover, the circadian rhythm of free cortisol was investigated on two days with 4 hourly samples obtained between 9 am and 9 pm. No differences were observed between smokers and nonsmokers in the circadian cortisol profiles.

In a second Study, 10 smokers and 17 nonsmokers participated in the same psychological stress test as in Study 1. In this experiment, only free cortisol was measured in saliva. Again, smokers showed significantly lower cortisol responses to stress compared to nonsmokers. We conclude that chronic nicotine consumption may lead to lower responses of multiple hormones not only to nicotine but to a variety of stimuli and that these alterations do not necessarily affect unstimulated circadian profiles of free cortisol.

(Supported by a grant from the *Forschungsrat für Rauchen und Gesundheit*.)

CONCURRENT MECAMYLAMINE/NICOTINE ADMINISTRATION

J. Rose[1,2], E. Levin,[2], F. Behm[2], E. Westman[1,3], R. Stein[1,2], J. Lane [2], and G. Ripka[2] [1]VA Medical Center, Durham, NC 27705, [2]Department of Psychiatry, Duke University, Durham, NC, [3]Department of Medicine, Duke University, Durham, NC

Co-administration of a nicotinic agonist with an antagonist may regulate receptor activation, thereby relieving withdrawal symptoms and blocking nicotine reward. In one study, 12 smokers rated the enjoyable effects of cigarette smoke after separate and combined administration of nicotine and the nicotinic antagonist mecamylamine. While each drug offset potential side effects of the other, they acted in unison to attenuate smoking reward. In a second study, 48 subjects participated in a randomized, double-blind, placebo-controlled smoking cessation trial. Nicotine skin patch therapy (21 mg/day for 6-8 weeks) + oral mecamylamine (2.5-5 mg b.i.d. for 5 weeks) was compared to nicotine patch + placebo. Mecamylamine treatment began two weeks before smoking cessation. Combined agonist-antagonist treatment produced significantly higher continuous smoking abstinence than agonist-alone treatment: 50% *vs* 16.7% at seven weeks (p=.015), 37.5% *vs* 12.5% at six months (p=.046) and 37.5% *vs* 4.2% at twelve months (p=.004). Concurrent nicotine-mecamylamine treatment has potential advantages over treatment using nicotine alone or mecamylamine alone.

EEG EFFECTS OF MECAMYLAMINE (MEC) IN CIGARETTE SMOKERS AND NONSMOKERS. W.B. Pickworth, M. Butschky and J.E. Henningfield. NIDA, Addiction Research Center, Baltimore, MD 21224.

MEC, a centrally acting nicotine antagonist, has proven useful in assessing the physiologic role of brain nicotine receptors. In a previous study, mecamylamine decreased the EEG effects of nicotine chewing gum in tobacco-deprived subjects and exacerbated the EEG signs of tobacco withdrawal by further increasing EEG theta power and decreasing alpha frequency. Those EEG effects could have been caused by antagonism of residual nicotine, or MEC could have altered a tonically active nicotinic cholinergic system involved in the regulation of CNS arousal. To clarify those alternative explanations six nonsmokers and five smokers participated in this outpatient study. In sessions separated by at least 48 hr, they were given 0, 5, 10 and 20 mg oral doses of MEC. EEG recordings, memory tests, and subjective questionnaires were administered before and up to 3 hr after drug. In the smokers, MEC caused dose-related increases in theta power and decreases in alpha frequency. The EEG effects in the nonsmokers were less pronounced; high doses decreased in theta power but alpha frequency was not decreased. These findings support the notion that nicotinic cholinergic systems partially regulate spontaneous EEG activity in smokers and nonsmokers. Dysregulation of these systems may be involved in the nicotine abstinence syndrome and in the pathophysiology of Alzheimers disease.

Section 14: Pharmacokinetics and pharmacology of metabolites

P 124

EFFECT OF NICOTINE AND COTININE ON NNK METABOLISM IN RATS. C. Kutzer, E. Richter and S.E. Atawodi. Walther Straub-Institute of Pharmacology and Toxicology, Ludwig-Maximilians-University, Nussbaumstr. 26, D-80336 Munich, Germany.

Recently, we have shown a significantly lower excretion of α-hydroxylation products of 4-(methylnitrosamino)-1-(3-pyridyl)-1-butanone (NNK) in the urine of rats co-administered a 500-fold higher dose of nicotine (Richter et al., Naunyn Schmiedebergs Arch Pharmacol 348:R167, 1993). In order to simulate the condition in smokers, the effect of nicotine and cotinine on the metabolism of NNK was studied at low chronic doses of [5-^{3}H]-NNK (4.2 nmol/kg/day) administered simultaneously with 6000-fold higher doses of the tobacco alkaloids (26 μmol/kg/day). The compounds were administered subcutaneously to male F344 rats weighing about 200 g for 28 days using Alzet minipumps. Total ^{3}H in urine and faeces was determined daily. The metabolites in urine were analysed by HPLC. The concentration of free and bound ^{3}H in blood was determined at the end of the study. Excretion of total ^{3}H reached a plateau within one week. After 28 days, the rats in all groups excreted about 77% of the cumulative dose, 65% in urine and 12% in faeces, respectively. The metabolite pattern in urine did not change over time. Co-administration of nicotine and cotinine did not decrease the urinary excretion of NNK α-hydroxylation products which accounted for 61.7±0.7%, 65.2±0.4% and 66.0±0.6% of total ^{3}H in the urine of control, nicotine- and cotinine-treated rats, respectively. The excretion of N-oxidation products accounted for 32-34% of total ^{3}H in the urine of all rats. However, hemoglobin binding of ^{3}H was significantly reduced from 16.3±2.1 pmol/g hemoglobin in control rats to 9.0±2.2 and 9.3±1.4 pmol/g in nicotine- and cotinine-treated rats, respectively, while the concentration of ^{3}H in plasma (0.5-0.6 pmol/ml) was not significantly affected by the tobacco alkaloids.

Supported by VERUM; S.E. Atawodi is an Alexander von Humboldt Research Fellow.

P 125

DISTRIBUTION AND RETENTION OF NICOTINE AND ITS MAJOR METABOLITE, COTININE, IN THE RAT AS A FUNCTION OF TIME. B.V. Rama Sastry, M.B. Chance, G. Singh, J.L. Horn, and V.E. Janson. Vanderbilt University Medical Center, Nashville, TN 37232-2125.

Nicotine (N) is oxidized to its major metabolite, cotinine (C), which has a long biological half-life (19-24 hr). The plasma concentration of C has been used as an index of tobacco-smoke exposure. It possibly increases the turnover rate of platelet activating factor (PAF) because it is a potent activator of PAF-hydrolase. It may play a significant role in tobacco-induced arterial thrombosis. Therefore, we studied the distribution and retention of N as it was metabolized to C in the rat. N (1 mg/kg, 5μCi/kg) was administered into the femoral vein of male Sprague Dawley rats under nembutal anesthesia. At different times (5-60 min) after N administration, N and its metabolite, C, in plasma, liver, kidney, heart, and brain were determined by HPLC (Fed. Proc. 46: 864, 1987). Within 5-10 min after administration, N concentrations reached peak values in plasma (2160 pmol/ml) as well as in brain. The plasma level of N decreased by 50% in 50 min (half-time). The half-time of N for brain was about 60 min. The half-times of N for the other organs were less than 20 min. The major metabolite, C, accumulated in plasma, and by about 30 min the concentration of N and C in plasma were about equal (890-1000 pmol/ml). While C accumulated in plasma, N was eliminated by the kidney. The concentration of N in kidney was 4 times higher than in plasma at 10 min, but twice higher at 60 min. At 60 min, C concentration in kidney was one-half of that in plasma. These observations indicate that N is renally eliminated or metabolized to C while C exhibits a long retention time and accumulates in plasma. Its accumulation in plasma may contribute to thrombosis. (Supported by HHS-NIH Grant DA-06207, The Council for Tobacco Research, USA, Inc., and The Study Center for Anesthesia Toxicology).

P 126

DETECTION OF A LONG-LIVED NICOTINE METABOLITE IN RAT BRAIN FOLLOWING PERIPHERAL NICOTINE ADMINISTRATION P.A. Crooks, M. Li and L.P. Dwoskin. College of Pharmacy, University of Kentucky, Lexington, KY, 40536 USA.

Little is known about the metabolic fate of nicotine in the CNS. Recent studies demonstrate the presence of cotinine and nicotine-N'-oxide in brain following peripheral nicotine administration. The present study clearly indicates that in addition to cotinine, a long-lived nicotine metabolite can be detected in brain. This metabolite is not nicotine-N'-oxide. S(-)[^{3}H-N-Methyl]-nicotine was administered peripherally (s.c.) to rats. Cation exchange high pressure liquid radiochromatography was utilized to separate and quantitate [^{3}H]nicotine and its [^{3}H]metabolites in plasma and brain homogenates at 5, 30, 60, 240, 1080 min following [^{3}H]nicotine administration. Results demonstrate that in brain the concentration of [^{3}H]nicotine peaked at 60 min after its administration and subsequently diminished to an undetectable concentration at 18 hrs. Two [^{3}H]metabolites of [^{3}H]nicotine were detected and quantified. One of these [^{3}H]metabolites had identical chromatographic characteristics to authentic cotinine. [^{3}H]Cotinine was not detected at 5 min after [^{3}H]nicotine administration, increased to peak concentrations at 4 hrs, and was moderately diminished by 18 hrs post administration. In contrast, [^{3}H]cotinine was a significant peripheral metabolite 5 min after [^{3}H]nicotine administration, and plateaued between 30 min to 4 hrs, with a significant reduction at 18 hrs. The second [^{3}H]metabolite observed in brain has not been identified as yet, but was highly polar and nonbasic in nature with characteristics similar to a glucuronidated biotransformation product. Concentrations of this [^{3}H]metabolite in brain were apparent at 5 min after [^{3}H]nicotine administration, increased over the first 30 min, and subsequently remained relatively constant over the 18 hr time period. This is the first report of the accumulation of a long-lived nicotine metabolite in the CNS after peripheral administration of nicotine. Studies are in progress to determine the identity of this new metabolite. (Supported by a grant from the Tobacco and Health Research Institute, Lexington, KY.)

INTER-INDIVIDUAL VARIATION OF NICOTINE UPTAKE AMONG SMOKERS

G. D. Byrd, J. H. Robinson, W. S. Caldwell, and D. J. deBethizy, R. J. Reynolds Tobacco Company, Product Evaluation Group, Winston-Salem, NC 27102.

Cigarette smokers have a wide variety of "tar" and nicotine yield products to choose from in the current market, ranging from 0.5 mg "tar" and less than 0.05 mg nicotine to 27 mg "tar" and 1.8 mg nicotine by FTC method. To better understand the relationship between FTC nicotine yields and actual nicotine uptake in smokers, we have studied nicotine uptake in 33 smokers that spanned four groups: 1 mg "tar" (1MG), ultra-low "tar" (ULT), full-flavor low "tar" (FFLT), and full flavor (FF) cigarette smokers. These cigarette categories had mean FTC nicotine yields of 0.14, 0.49, 0.67 and 1.13 mg/cigarette, respectively. The subjects smoked *ad libitum* their usual brand of cigarette and submitted a 24 h urine sample for total nicotine uptake analysis over a period where the number of cigarettes smoked was recorded. Nicotine uptake was determined by monitoring urinary nicotine and its metabolites, including the glucuronide conjugates. Daily nicotine uptake was 9.1 ± 7.3 mg (range 1-21 mg) for 1MG, 19.2 ± 10.0 mg (range 4-42 mg) for ULT, 21.8 ± 9.4 mg (range 13-38 mg) for FFLT, and 37.1 ± 14.4 mg (range 21-60 mg) for FF smokers; on a per cigarette basis this equated to 0.23 ± 0.11, 0.56 ± 0.23, 0.60 ± 0.18, and 1.19 ± 0.43 mg nicotine, respectively. Although the number of subjects in each group is limited, means for the different groups showed that lower FTC yield cigarettes result in not only less nicotine uptake per 24 h period, but also per cigarette smoked. These data suggest that nicotine uptake is a function of individual smoking behavior within product design limits. We conclude from these data that, while FTC yield cannot precisely predict nicotine uptake for an individual smoker, it is useful in predicting and comparing actual nicotine uptake by smokers who select cigarettes with a particular FTC yield.

URINARY EXCRETION OF MINOR TOBACCO ALKALOIDS BY SMOKERS AND SMOKELESS TOBACCO USERS. BIOCHEMICAL MARKERS FOR TOBACCO USE IN PERSONS USING NICOTINE-CONTAINING MEDICATIONS.

Peyton Jacob, III, Neal L. Benowitz, Herb Severson, and Dorothy Hatsukami. University of California, San Francisco, CA; Pacific Research Institute, Eugene, OR; University, of Minnesota, Minneapolis, MN, USA.

Tobacco and tobacco smoke contain various alkaloids structurally related to nicotine. Some of these alkaloids are of interest due to their pharmacologic activity, because they are precursors to carcinogenic nitrosamines, or because they are potentially useful markers for tobacco use. Because they are not metabolically derived from nicotine, the alkaloids anabasine and anatabine may differentiate tobacco use from medication with nicotine. Concentrations of nicotine, cotinine, nornicotine, anabasine, and anatabine in urine of 99 cigarette smokers averaged 1960, 1790, 113, 21, and 22 ng/ml, respectively. Corresponding concentrations in urine of 215 smokeless tobacco (snuff or chewing tobacco) users averaged 1460, 2380, 119, 25, and 44 ng/ml. The higher concentrations of anatabine in urine of smokeless tobacco users is presumably due to absence of pyrolysis of this alkaloid that occurs during smoking. This suggested that anatabine levels may be used to distinguish smokers from smokeless tobacco users, and it was found that the ratio of anatabine to cotinine in urine could be used to distinguish the two groups. In two studies employing nicotine gum to aid smokeless tobacco cessation, the absence of anabasine and anatabine was highly correlated with self reported tobacco cessation. Consequently, anabasine and anatabine may be used to confirm cessation of tobacco use in clinical trials involving nicotine replacement.

BIOMONITORING OF TOBACCO-SPECIFIC NITROSAMINES IN URINE.

A.R.Tricker, M.Meger, G.Scherer, F.Adlkofer, A.Pachinger[+] and H.Klus[+]. Analytisch-biolog-isches Forschungslabor Prof. F.Adlkofer, Goethestrasse 20, 80336 Munich, Germany, and [+]Ökolab, Gesellschaft für Umweltanalytik, Hasnerstrasse 124a, A-1160 Vienna, Austria.

Nicotine is a potential precursor to N-nitroso compounds present in tobacco and tobacco smoke. Biomonitoring methods have been developed for the determination of 4-(N-methylnitrosamino)-1-(3-pyridyl)-1-butanone (NNK) metabolites and 4-(N-methylnitrosamino)-4-(3-pyridyl)butyric acid (Iso-NNAC) in urine. NNK is partially metabolized by carbonyl reduction to 4-(N-methylnitrosamino)-1-(3-pyridyl)-1-butanol (NNAL) which is excreted in urine together with its glucuronide conjugate [4-(N-methylnitrosamino)-1-(3-pyridyl)but-1-yl]-β-O-D-glucosiduronic acid (NNAL-Gluc). Iso-NNAC is excreted almost quantitatively in urine and feces. Urine samples from smokers (n = 13) contained 229 ± 176 (range 40-667) ng NNAL/day, 425 ± 325 (range 31-1053) ng NNAL-Gluc/day and 23.2 ± 49.4 (range 0-163) ng Iso-NNAC/day. NNAL, NNAL-Gluc and Iso-NNAC were not detected in urine from nonsmokers (n = 10) and after administration of nicotine (oral: 12-40 mg; transdermal: 21 mg) or cotinine (oral: 60 mg), with or without oral nitrate supplementation (3 x 50 mg NO_3^-), to smokers after 1 week of smoking abstinence. It is concluded that the presence of tobacco-specific nitrosamines in urine is due only to exogenous exposure to NNK and Iso-NNAC from active smoking and not from endogenous nitrosation of nicotine and/or its metabolites.

P 130

NICOTINE IMINIUM IONS ARE NOT DETECTED IN SMOKERS' URINE. W. S. Caldwell, G. D. Byrd, G. P. Dobson, and G. M. Dull. Research and Development, R. J. Reynolds Tobacco Co., Winston-Salem, NC, 27102

The major pathway for the metabolism of nicotine in humans involves cytochrome P-450 catalyzed oxidation of the 5' position of the pyrrolidine ring to form nicotine-1',5'-iminium ion. Pseudooxynicotine (PON), a major metabolite of nicotine in certain soil bacteria and fungi, exists in equilibrium with nicotine-1'-2'-iminium ion in aqueous solution. A recent report that both nicotine-1',5'-iminium ion and nicotine-1'-2'-iminium ion were detected in smokers' urine (Neurath *et al.* (1992) *Med. Sci. Res.* **20**, 853-858) led to speculation that PON may also be a human metabolite of nicotine. Using proton and carbon NMR spectroscopy, we elucidated the solution chemistry of PON. This information was used to develop a method for the quantitative determination of PON and nicotine-1',5'-iminium ion in aqueous solutions. Aqueous samples were spiked with the internal standard methyl-d_3-PON, brought to pH 10 with saturated potassium carbonate, treated with an excess of KCN, and extracted with methylene chloride. The organic fraction was analyzed by GC/MS using a DB5 capillary colum. Limits of detection for the two analytes were approximately 5 ng/mL. We found that unless the samples were kept in the dark, this method was subject to artifactual formation of the iminium ions due to photochemical oxidation of nicotine. When smokers' urine and culture medium resulting from the incubation of nicotine with fresh human hepatocytes were assayed, no PON nor nicotine-1',5'-iminium ion were detected. Earlier reports of PON and nicotine-1',5'-iminium ion in smokers' urine may have fallen victim to artifact formation.

P 131

DIRECT DETERMINATION OF NICOTINE-N-GLUCURONIDE IN HUMAN BIOLOGICAL SAMPLES. G. D. Byrd and W. S. Caldwell, Biological Chemistry Division, R. J. Reynolds Tobacco Co., Winston-Salem, NC 27102, P. A. Crooks, A. Ravard, and B. S. Bhatti, College of Pharmacy, University of Kentucky, Lexington, KY 40536-0082.

An important part of nicotine metabolism in humans occurs through phase II reactions that produce glucuronide conjugates of nicotine and its two major metabolites, *trans*-3'-hydroxycotinine and cotinine, as proven by their occurrence in the urine of smokers and snuff users. Most of the evidence for these glucuronides has been by indirect methods using either enzymatic (ß-glucuronidase) or alkaline hydrolysis of biological samples to release the aglycons which are then detected by various analytical means. We report here direct evidence that nicotine glucuronide is produced in humans as the quarternary N-glucuronide with linkage through the pyridine nitrogen. Florisil solid phase extraction was used to partially purify the material from urine. Detection of the glucuronide was by thermospray-LC/MS. The identified compound had the same retention time as a synthetic standard and gave the same mass spectrum characterized by the protonated molecular ion and the protonated aglycon. Using $^2H_3$-labeled synthetic standard, nicotine-N-glucuronide was determined in urine samples from 6 smokers. Concentrations ranged from 2.2 to 7.6 nmol/mL with a limit of detection of 1.3 nmol/mL. The average amount of total nicotine excreted by the six smokers as nicotine-N-glucuronide was 6.26 ± 4.2% with a range of 0.9 to 14.3%. Large interindividual variability (>10 fold) such as this results from differences among smokers in nicotine uptake and metabolism and has been observed in other studies. Nicotine-N-glucuronide was also identified as a metabolic product of nicotine when incubated with human hepatocytes. This work confirms nicotine-N-glucuronide as an important phase II nicotine metabolite in humans.

P 132

NICOTINE METABOLISM AND *CYP2D6* POLYMORPHISM IN A POPULATION OF NON-TOBACCO USERS. S Cholerton, A Arpanahi, NW McCracken, C Boustead, H Taber, E Johnstone, J Leathart, AK Daly and JR Idle. Pharmacogenetics Research Unit, Department of Pharmacological Sciences, University of Newcastle upon Tyne, Newcastle upon Tyne, NE2 4HH, UK.

In man the major primary C-oxidation product of nicotine (N) metabolism is cotinine (C). *In vitro* studies suggest that several cytochromes P450 (CYP) are capable of mediating the C-oxidation of nicotine, one of which is the polymorphically expressed CYP2D6. The aim of this study (which had ethical approval) was to determine the influence of *CYP2D6* genotype on the metabolism of nicotine *in vivo* in unrelated, non-tobacco users. One hundred and twenty four (80 female; 19-49 years) volunteers were given oral nicotine (2mg) and urine was alkalinized (≥pH7). The presence of the *CYP2D6* wild-type (wt) allele and *CYP2D6A*, *CYP2D6B*, *CYP2D6D* and *CYP2D6E* mutant alleles was determined in 118. The 0-24h urinary recovery (% dose) of N [range (median)], C, 3-hydroxycotinine and 3'-hydroxycotinine glucuronides was 0.04-12.6 (0.8), 1.8-28.4 (5.5), 0-71.4 (8.0) and 0-25.4 (0.8) respectively. The N/C ratio (mean±se) for homozygotes for *CYP2D6* wt (n=62), heterozygotes for a *CYP2D6* inactivating mutation (n=43) and homozygotes for a *CYP2D6* inactivating mutation (n=13) was 0.20±0.03, 0.19±0.03 and 0.63±0.14 respectively. Thus homozygous mutants have significantly poorer nicotine to cotinine metabolism than both homozygous wt (P=0.005) and heterozygotes (P=0.005) as judged by the Mann-Whitney U-test.

Supported by a grant from Smokeless Tobacco Research Council Inc., USA

INHIBITION OF THE METABOLISM OF NICOTINE TO COTININE IN HUMAN LIVER MICROSOMES BY QUINIDINE AND COUMARIN. S Cholerton, NW McCracken and JR Idle. Pharmacogenetics Research Unit, Department of Pharmacological Sciences, University of Newcastle upon Tyne, Newcastle upon Tyne, NE2 4HH, UK.

Previous *in vitro* studies using cytochrome P450 (CYP) expression systems suggest that several isozymes, including CYP2D6 and CYP2A6, are capable of mediating the *C*-oxidation of nicotine to cotinine. The aim of this study was to determine the relative contributions of CYP2D6 and CYP2A6 to the *C*-oxidation of nicotine in human liver microsomes obtained from 6 human livers. Their ability to convert nicotine to cotinine ranged from 0.109-0.835 nmoles/min/mg protein and to 7-hydroxylate coumarin (an index of CYP2A6 activity) ranged from 0.182-0.779 nmoles/min/mg protein. There was good correlation between the rate of cotinine formation and coumarin 7-hydroxylase activity (R^2=0.646). In the presence of quinidine (Q), a selective CYP2D6 inhibitor, cotinine formation was inhibited in 5 of the 6 livers (16-92%). Coumarin (C) inhibited the formation of cotinine in 4 livers (29-66%). In the presence of Q+C, cotinine formation was inhibited in all livers (26-96%). In contrast troleandomycin (T), a selective inhibitor of CYP3A4-catalyzed reactions, had no significant effect in any preparation. The results suggest that CYP2D6 and CYP2A6 have a significant role in the metabolism of nicotine to cotinine and that CYP3A4 has very little involvement in this metabolic reaction. Thus it is likely that expression of CYP2D6 and CYP2A6 will influence an individuals ability to metabolize nicotine to cotinine.

Supported by a grant from Smokeless Tobacco Research Council Inc., USA

SIMULTANEOUS DETERMINATION OF NICOTINE AND COTININE IN PLASMA USING CAPILLARY COLUMN GAS CHROMATOGRAPHY WITH NITROGEN-SENSITIVE DETECTOR. K. Yan and J.G. Besner*, Faculty of pharmacy, University of Montreal, Montreal, Canada, H3C 3J7.

A rapid and sensitive method to determine plasma nicotine and cotinine simultaneously is described. After a simple one-step solvent extraction procedure with dichloromethane in alkalinized plasma, the quantitative analyses are performed by capillary column (DB-5, 0.53 mm X 30 m X 0.88 μm) gas chromatography using a nitrogen-sensitive detector. The structural analogues of nicotine and cotinine, ethyl-nornicotine (20 ng/ml) and 5-methyl cotinine (200 ng/ml), are used as internal standards respectively. Other special measures to avoid contamination from external sources such as atmosphere, solvents and laboratory equipment, which constitutes the major limiting factor of nicotine assay, are also undertaken. The method has been demonstrated to detect nicotine and cotinine plasma concentration within 1 and 8 ng/ml, respectively. The overall recovery for both drugs are more than 96 %. The precision of the method has been investigated by determining the reproducibility of quality control samples at different levels of nicotine and cotinine within the working ranges. The average coefficient of variation over the nicotine range 1-100 ng/ml is 5 % and for cotinine over range 20 to 900 ng/ml is 3.6 %. The method described is allowed to be used for measurement of these compounds in both smokers and non-smokers after the use of nicotine patch.

SIMULTANEOUS ANALYSIS OF NICOTINE (N) AND COTININE (C) IN HUMAN PLASMA BY REVERSED PHASE HPLC USING ULTRAVIOLET AND ELECTROCHEMICAL DETECTION. M. Bouhajib and J.G. Besner. Faculty of Pharmacy, University of Montreal, Montreal, Canada, H3C 3J7.

Analysis of N and C in biological matrix after transdermal administration of N is a challenge due to the low levels of N measured, lack of sensitivity of UV detection for N and possible N contaminations during the analytical procedures. Aliquots of blank plasma (1 ml) spiked with N and C concentrations varying respectively between 0-50 ng/ml and 0-250 ng/ml together with 70 ng of N-ethyl Nornicotine and 100 ng of 5-methyl cotinine as internal standards are extracted for 10 min with 3 ml of chloroform ethylacetate (90:10). After centrifugation the organic phase is back extracted with 0.3 ml of 0.05 N H_2SO_4. 30 mcl are injected into a 25 cm HPLC column packed with S-5CN. The mobile phase composed of acetonitrile: methanol: acetic acid: 20 mM dibutylamine in water (3.5: 3: 0.6: 92.9) is set at 1 ml/min. The detection is as follows: first a UV detector set at 259 nm measures C while an ESA coulometer operated at +0.95 V detects N. The four substances are eluted within 9-14 min and are devoid of any endogenous interference. The precision and accuracy are greater than 95 % for both drugs and the limits of detection are respectively 1 ng/ml of N and 5 ng/ml of C. The overal recovery is greater than 87.5 % for any of the 4 analytes. This method is excellent for the pharmacokinetic evaluation of N and C in human subjects treated with nicotine transdermal therapeutic system.

P 136

THE GENOTOXIC POTENTIAL OF NICOTINE AND ITS MAJOR METABOLITES.
D. Doolittle, C. Fulp, C. Lee, W. Caldwell and D. deBethizy. Research and
Development; R. J. Reynolds Tobacco Company, Winston-Salem, North Carolina
27102, U.S.A.

Nicotine is a naturally occurring alkaloid found in members of the solanaceous plant family,
which includes tobacco. Nicotine is metabolized by mammalian enzymes, primarily
cytochrome P450s. Studies on the genotoxic potential of these metabolites are limited.
Nicotine and four of its major metabolites: cotinine, nicotine N'-oxide, cotinine N-oxide, and
3'-hydroxycotinine were evaluated for genotoxic potential in the Salmonella mutagenicity
assay (strains TA 98, TA 100, TA 1535, TA 1537, and TA 1538) at concentrations ranging
from 0 to 1000 µg/plate and in the Chinese hamster ovary sister chromatid exchange (SCE)
assay at concentrations ranging from 0 to 1000 µg/ml. All assays were conducted with and
without S9 metabolic activation. None of the five compounds increased the frequency of
mutations or the frequency of SCEs. These results indicate that nicotine and its major
metabolites are not genotoxic.

P 137

**NICOTINE AND COTININE INHIBIT THE MUTAGENICITY OF N-NITROSAMINES
PRESENT IN TOBACCO SMOKE.** C. Lee, C. Fulp, E. Bombick, and D. Doolittle.
Research and Development; R. J. Reynolds Tobacco Company, Winston-Salem,
North Carolina, 27102 U.S.A.

Nicotine and N-nitrosamines are both found in tobacco smoke, with the concentration of
nicotine several orders of magnitude higher than that of nitrosamines. The major genotoxic N-
nitrosamines present in tobacco smoke are N-nitrosodimethylamine (NDMA), 4-
(methylnitrosamino)-1-(3-pyridyl)-1-butanone (NNK) and N'-nitrosonornicotine (NNN). These
nitrosamines require metabolic activation by cytochrome P-450's for the expression of
mutagenicity. Although nicotine has been shown to inhibit the metabolic activation of NNK, its
effect on the mutagenicity of N-nitrosamines has not been reported. The ability of nicotine and
its metabolites to inhibit the mutagenicity of tobacco-related N-nitrosamines was tested on
Salmonella typhimurium strain TA1535 and TA100 in the presence of an exogenous metabolic
activation system (S9). Nicotine and cotinine inhibited the mutagenicity of NDMA and NNK, but
not NNN, in a dose-dependent manner. Results of cell survival tests on nicotine and cotinine
ruled out the possibility that the reduction of mutagenicity was due to cell killing. The induction
of SCE in CHO cells by NNK (0.15 - 1.16 mM) in the presence of metabolic activation was also
significantly reduced by nicotine (6.2 mM) and cotinine (11.4 mM), indicating that both nicotine
and cotinine inhibit the genotoxicity of NNK in a mammalian system. These results demonstrate
that nicotine and its primary metabolite, cotinine, act as antimutagens for some of the major
genotoxic N-nitrosamines in tobacco smoke.

Author Index

EXPERIENTIA SUPPLEMENTUM

Nonselective Cation Channels
Pharmacology, Physiology and Biophysics

Edited by
D. Siemen, *University of Regensburg, Regensburg, Germany*
J. Hescheler, *Free University of Berlin, Berlin, Germany*

1993. 304 pages. Hardcover. ISBN 3-7643-2888-6 (EXS 66)

In 1981, Neher and Sakmann published a pioneering paper describing the patch clamp method for measuring the currents flowing through individual ion channels. Apart from channels which are selective for a single species of ion, numerous "nonselective channels" have been found since then, which are activated by various agonists and differ in selectivity and amino acid sequence. Today it is widely acknowledged that these channels belong to the most important components of the cell membrane and fulfil a wide spectrum of different functions.

Nonselective Cation Channels is the first book to report on the immense variety and diversity of nonspecific ion channels, ranging from the nicotinic acetylcholine receptor to the gap junction channel. Written by recognized international experts, its coverage also includes the role of nonselective cation channels in cardiac and smooth muscle cell functions, and their importance in human platelets and endothelial cells. Sections on receptor-activated and mechanically sensitive cation channels are also included.

Please order through your bookseller
or directly from:
Birkhäuser Verlag AG, P.O. Box 133
CH-4010 Basel / Switzerland
Fax ++41 / 61 721 79 50
E-Mail: 100010.23@compuserve.com

Orders from the USA or Canada
should be sent to:
Birkhäuser Boston, 333 Meadowlands Parkway,
Secaucus, NJ 07094-2491 / USA
Call Toll-Free 1-800-777-4643

For more information on recent and forthcoming books and journals you can order the BIRKHÄUSER LIFE SCIENCES BULLETIN, published twice a year and free of charge.

Birkhäuser Verlag • Basel • Boston • Berlin

Fractals in Biology and Medicine

Edited by
T.F. Nonnenmacher, *Mathematische Physik, Universität Ulm, Germany*
G.A. Losa, *Instituto Cantonale di Patologia, Locarno, Switzerland*
E.R. Weibel, *Anatomisches Institut Universität Bern, Switzerland*

1994. 397 pages. Hardcover. ISBN 3-7643-2989-0

With a chapter by B. Mandelbrot

Fractals in Biology and Medicine explores the potential of fractal geometry for describing and understanding biological organisms, their development and growth as well as their structural design and functional properties. It extends these notions to assess changes associated with disease in the hope to contribute to the understanding of pathogenetic processes in medicine.

The book is the first comprehensive presentation of the importance of the new concept of fractal geometry for biological and medical sciences. It collates in a logical sequence extended papers based on invited lectures and free communications presented at a symposium in Ascona, Switzerland, attended by leading scientists in this field, among them the originator of fractal geometry, Benoât Mandelbrot.

Fractals in Biology and Medicine begins by asking how the theoretical construct of fractal geometry can be applied to biomedical sciences and then addresses the role of fractals in the design and morphogenesis of biological organisms as well as in molecular and cell biology. The consideration of fractal structure in understanding metabolic functions and pathological changes is a particularly promising avenue for future research.

Please order through your bookseller
or directly from:
Birkhäuser Verlag AG, P.O. Box 133
CH-4010 Basel / Switzerland
Fax ++41 / 61 721 79 50
E-Mail: 100010.23@compuserve.com

Orders from the USA or Canada
should be sent to:
Birkhäuser Boston, 333 Meadowlands Parkway,
Secaucus, NJ 07094-2491 / USA
Call Toll-Free 1-800-777-4643

For more information on recent and forthcoming books and journals you can order the BIRKHÄUSER LIFE SCIENCES BULLETIN, published twice a year and free of charge.

Birkhäuser Verlag • Basel • Boston • Berlin